Computational Molecular Science and Engineering Forum 2022

Held at the 2022 AIChE Annual Meeting

Phoenix, Arizona, USA
13-18 November 2022

ISBN: 978-1-7138-7905-3

Printed from e-media with permission by:

Curran Associates, Inc.
57 Morehouse Lane
Red Hook, NY 12571

Some format issues inherent in the e-media version may also appear in this print version.

Copyright© (2022) by AIChE
All rights reserved.

Printed with permission by Curran Associates, Inc. (2023)

For permission requests, please contact AIChE
at the address below.

AIChE
120 Wall Street, FL 23
New York, NY 10005-4020

Phone: (800) 242-4363
Fax: (203) 775-5177

www.aiche.org

Additional copies of this publication are available from:

Curran Associates, Inc.
57 Morehouse Lane
Red Hook, NY 12571 USA
Phone: 845-758-0400
Fax: 845-758-2633
Email: curran@proceedings.com
Web: www.proceedings.com

TABLE OF CONTENTS

MACHINE LEARNING FOR SOFT MATERIALS I

HONORARY SESSIONS FOR KEITH GUBBINS' 85TH BIRTHDAY (INVITED ONLY)

PLENARY SESSION: COMPUTATIONAL MOLECULAR SCIENCE AND ENGINEERING FORUM (INVITED TALKS)

PRACTICAL APPLICATIONS OF COMPUTATIONAL CHEMISTRY AND MOLECULAR SIMULATION FOR COMPLEX FLUIDS AND PHASE BEHAVIOR

RECENT ADVANCES IN MOLECULAR SIMULATION METHODS

RECENT ADVANCES IN MULTISCALE METHODOLOGIES

SOFTWARE ENGINEERING IN AND FOR THE MOLECULAR SCIENCES

SPOTLIGHTS IN THERMODYNAMICS AND COMPUTATIONAL MOLECULAR SCIENCE (INVITED TALKS)

THE INDUSTRIAL FLUID PROPERTIES SIMULATION CHALLENGE

APPLICATIONS OF MOLECULAR MODELING TO STUDY INTERFACIAL PHENOMENA II

MACHINE LEARNING FOR SOFT MATERIALS II

PRACTICAL APPLICATIONS OF COMPUTATIONAL CHEMISTRY AND MOLECULAR SIMULATION FOR SOLVENTS AND INORGANIC MATERIALS

RECENT ADVANCES IN MOLECULAR MODELING OF INTERFACIAL THERMODYNAMICS AND DYNAMICS

PRACTICAL APPLICATIONS OF COMPUTATIONAL CHEMISTRY AND MOLECULAR SIMULATION FOR POLYMERS AND BIOLOGICAL SYSTEMS

9a Atomistic-Level Insights into the Interfacial Interactions between Amyloid-Beta and Phospholipid Bilayers for the Treatment of Alzheimer's Disease

Bailey Zinger, Joel L. Kaar, Kayla Sprenger

Alzheimer's Disease (AD) is a devastating neurological disease that affects 6.5 million Americans today, and AD diagnoses are projected to more than double by 2050 [1]. A hallmark of AD is the aggregation of misfolded amyloid-β (Aβ) proteins in the brain into Aβ oligomers (AβO), thought to be mediated by interfacial interactions between Aβ and host cell membranes [2]. Experimental data have shown AβO are soluble, neurotoxic, and contribute to cognitive decline in people with AD [3]. New treatment modalities aimed at preventing or reversing Aβ aggregation by controlling these interactions offer promise for combatting AD. Recently, Sosa et al. discovered that lipid vesicles—spherical structures composed of a cell membrane-like phospholipid bilayer, surrounding an aqueous interior—could prevent aggregation of misfolded Aβ and catalyze the breakdown of aggregated Aβ species like AβO [4]. However, the mechanisms behind these promising results remain entirely unclear.

It is hypothesized that the binding energy between AβO and the bilayer must not be too strong such that AβO will be overly stabilized on the surface and unable to undergo conformational changes, but not too weak such that AβO will desorb from the surface without degrading or refolding into a non-neurotoxic form. We explored this hypothesis via molecular dynamics (MD) simulations of the different forms of Aβ (monomers, AβO, and fibrils) adsorbing onto bilayers with lipid compositions mimicking the experiments of Sosa et al. Furthermore, we employed the enhanced sampling method metadynamics combined with temperature-based replica exchange (PTMetaD-WTE; [5]) to ensure robust sampling of the surface-bound conformations of Aβ in the timescales accessible to our simulations. Our results provide new insights into how biological membranes catalyze the degradation of neurotoxic Aβ species in the brain. This knowledge could possibly lead to the design of new and/or improved breakthrough treatments for combatting AD, thereby impacting millions of lives worldwide.

References:

[1] Alzheimer's facts and figures report | *Alzheimer's Association*. https://www.alz.org/alzheimers-dementia/facts-figures.

[2] Khondker A, et al. "Membrane-accelerated amyloid-β aggregation and formation of cross-β sheets", *Membranes*, **2017**, 7, 49.

[3] Gibbs E, et al. "A rationally designed humanized antibody selective for amyloid beta oligomers in Alzheimer's Disease", *Sci Rep*, **2019**, 9, 9870.

[4] Chaparro-Sosa AF, et al. "Mixed phospholipid vesicles catalytically inhibit and reverse amyloid fibril formation", *J Phys Chem Lett*, **2020**, 11, 7417-7422.

[5] Deighan M, et al. "Efficient simulation of explicitly solvated proteins in the well-tempered ensemble", *J Chem Theory Comput*, **2012**, 8, 2189-2192.

9c Computational Discovery of Plastic-Binding Peptides for Microplastic Remediation

Michael Bergman, Xingqing Xiao, Carol Hall

Plastic pollution on the micrometer or nanometer scale, termed microplastics or nanoplastics (MNPs), is a pressing environmental issue. The exact amount of MNP pollution in the environment is unknown, but it has been estimated that there are 10 to 20 million tons in the top 200 meters of the Atlantic Ocean alone. MNPs have also been found in sea ice, the atmosphere, and food and water sources. The ubiquitous presence of MNPs causes many organisms, including humans, to consume MNPs daily. MNP consumption has been suggested to harm organisms due to toxic species adsorbed on the MNP surface. The environmental and health risks associated with MNPs have created a need to remove MNP pollution from the environment.

Our goal is to computationally design plastic-binding peptides as part of a larger group effort to remediate MNP pollution in bodies of water. The designed peptides will be conjugated to active particle microcleaners to help capture MNPs, conjugated to liquid crystal sensors to create a platform for characterizing MNP waste, and expressed on the surface of engineered microbes to increase cell adhesion to MNPs and promote subsequent MNP degradation. Peptides are being designed using the **Pep**tide **B**inding **D**esign (PepBD) algorithm developed in the Hall lab, which uses Metropolis Monte Carlo sampling to search for peptide sequences with high affinity for a target molecule. We aim to design peptide binders for polyethylene, polypropylene, and polystyrene, three of the most-commonly produced plastics.

We report the results for our initial plastic-binding peptide designs. We began by verifying that Amber's GAFF2 parameters combined with partial charges calculated using *ab initio* methods reproduce experimental values of the density, radius of gyration, and heat capacity of the three target plastics. We then created amorphous and crystalline plastic surfaces that were subsequently used in molecular dynamics adsorption simulations of peptides known to have affinity for plastic. The adsorbed configurations of the peptides were used as starting structures for PepBD to search for other peptides that adsorb strongly to plastics. The peptides with the best scores underwent traditional molecular dynamics, well-tempered metadynamics, and steered molecular dynamics simulations to determine the interaction energies, adsorption free energies, adsorbed conformations, and desorption forces of the peptides. We found a polyethylene-binding peptide that has a free energy of adsorption and average pull-off force that is similar to that of a polyethylene-binding peptide found by others through phage display. This suggests we can computationally discover peptides with affinities that are equivalent to those of peptides discovered using phage display, thereby reducing the cost and time of finding plastic-binding peptides and potentially allowing for the discovery of peptides with even greater affinity to plastic than the peptides that are currently known. We are working with collaborators to experimentally test the computationally discovered peptide using atomic force microscopy.

9d Computational Investigation of Bijels As Separation Systems

Marco Tulio Portella, Xuan Duy Thao Nguyen, Dimitrios Papavassiliou

Bicontinuous Interfacially Jammed Emulsion Gels, also known as *Bijels*, find many applications in the industry including catalysis, separation processes and tissue engineering [1]. Recently, they have been considered as options for reactive separation media due to their large interfacial areas and ease of removal and insertion of reagents and products. In this work, *Bijel* behavior are investigated through the use of Dissipative Particle Dynamics (DPD) computations, which is available through the Large-scale Atomic/Molecular Massively Parallel Simulator (LAMMPS) software package[2-5]. The two fluid phases are water and diethyl phthalate, stabilized with silica nanoparticles. Surface parameters, such as interfacial tension and contact angle, are measured using the DPD coarse grained simulations in order to validate the computations. Thereby, the interactions between the immiscible phases and the jammed nanoparticles can be evaluated, with particular interest in interfacial mass transfer. With knowledge of interphase transport properties, the interfacial diffusion phenomena will be discussed. Thereafter, the migration of lipophilic compounds from the water to the oil phase will be simulated to explore the use of bijels as reactive separtation media.

ACKNOWLEDGEMENTS

The support of NSF under grant EFRI-2132141 is gratefully acknowledged as the use of computing facilities at the University of Oklahoma Supercomputing Center for Education and Research (OSCER) and at XSEDE (under allocation CTS-090025).

REFERENCES

1. Cha, S., Lim, H. G., Haase, M. F., Stebe, K. J., Jung, G. Y., Lee, D. "Bicontinuous Interfacially Jammed Emulsion Gels (bijels) as Media for Enabling Enzymatic Reactive Separation of a Highly Water Insoluble Substrate." *Scientific Reports* 2019, **9** (1), 6363.
2. Groot, R.D., Warren, P.B. "Dissipative particle dynamics: Bridging the gap between atomistic and mesoscopic simulation." *The Journal of Chemical Physics* 1997, **107**(11), 4423-4435.
3. Plimpton, S., "Fast Parallel Algorithms for Short-Range Molecular Dynamics." *Journal of Computational Physics* 1995, **117**(1), 1-19.
4. Nguyen, T.X.D., Vu, T.V., Razavi, S., Papavassiliou, D.V., "Coarse Grained Modeling of Multiphase Flows with Surfactants," *Polymers* 2022 **14**(3) Art 543 (19 pages).
5. Vu, T.V., Papavassiliou, D.V., "Synergistic effects of surfactants and heterogeneous nanoparticles at oil-water interface: Insights from computations," *Colloid andamp; Interf. Sci.* 2019, **553,** 50-58.

9e Adsorption of Trace Metal Contaminants from Coal-Derived Syngas on Metal Surfaces and Novel Bimetallic Adsorbents

Dwijraj Mhatre, Divesh Bhatia

To produce methanol from syngas obtained from the gasification of coal, the syngas needs to be free from trace metal contaminants such as elemental gas-phase mercury (Hg^0), which can poison the methanol synthesis catalyst. The temperature at the gasifier outlet and those required for the synthesis of methanol are high (>200 °C). However, most of the commercial adsorbents are inefficient for Hg^0 removal from syngas at high temperatures and in the presence of water vapor. Previous studies have shown that the strongest Hg^0 adsorption occurs on expensive Pd-based adsorbents, and the efficiency of these adsorbents drastically decreases with an increase in temperature. In this work, novel bimetallic adsorbents are identified for the removal of Hg^0 at high temperatures, and insights into the segregation behavior of alloys and the poisoning effect of Hg^0 is obtained. Further, the most promising monometallic and bimetallic adsorbents are synthesized and their efficacy for Hg^0 adsorption at high temperatures is shown.

DFT calculations are coupled with Gibbs free energy calculations to evaluate the feasibility of Hg^0 adsorption on two different types of bimetallic adsorbents, i.e., homogenous alloys and overlayers, for Hg^0 removal at temperatures between 50 to 200 °C. Hg^0 binding energy and Gibbs free energy calculations are performed on 56 different compositions of bimetallic adsorbents. For most homogeneous alloys and over-layered adsorbents, Hg^0 adsorption is stronger than that on the individual component metals due to the ligand and strain effects. Novel bimetallic adsorbents comprising of Rh, Ru, or Pd overlayers on Ag and Au exhibit a high Hg^0 adsorption strength, and the binding is even stronger than that on the best reported material, i.e., monometallic Pd. The effect of increasing overlayers of Pd, Pt, Rh, Ir, and Ru is evaluated, and it is found that even a single overlayer of these metals on Ag and Au increases the binding strength significantly. The d-band center, d-band width, and upper edge of the d-band for various monometallic and bimetallic surfaces are calculated. An increase in the Hg^0 binding strength is correlated to the upward shift in the d-band center (Figure 1 (a)). Further, segregation energy calculations predict the tendency of Hg to segregate towards the surface of amalgams and disturb the perfect planar geometry of the Pd, Pt, Rh, Ru, Ir, Cu, Ag, and Au surfaces to form a non-crystalline Hg-rich amalgam surface. An analysis of the binding of various adsorbates (H, O, N, and S) shows that the adsorption becomes significantly weaker on various sites in close proximity to pre-adsorbed Hg. Moreover, for certain adsorbates, adsorbents and site locations, the adsorption does not take place on the proximal sites. The partial density of states is analyzed for various amalgam surfaces, which explains the poisoning effect of Hg^0 on metallic catalysts.

The bimetallic adsorbents shortlisted from DFT and Gibbs free energy calculations are prepared by the galvanic displacement method, whereas the monometallic adsorbents are prepared using the incipient wetness method. The initial Hg^0 adsorption efficiency of various lab-synthesized monometallic and bimetallic adsorbents supported on doped aluminas (SIRAL-40, SASOL) and mesoporous SBA-15 is shown in Figure 1 (b). Ag supported on SIRAL-40 (S-40) is efficient only at low temperatures (less than 150 °C). However, the bimetallic adsorbents Pd-Ag/S-40 and Rh-Ag/S-40 exhibit a significantly high initial adsorption efficiency (>96%) at 300 °C. Moreover, these adsorbents are active for Hg^0 adsorption even at 350 °C. To the best of our knowledge, any material capable of adsorbing Hg^0 at such high temperatures is not reported in the literature, thus marking a breakthrough in warm syngas cleanup. More bimetallic adsorbents identified by DFT will be prepared and the effect of operating conditions will be described in the full manuscript.

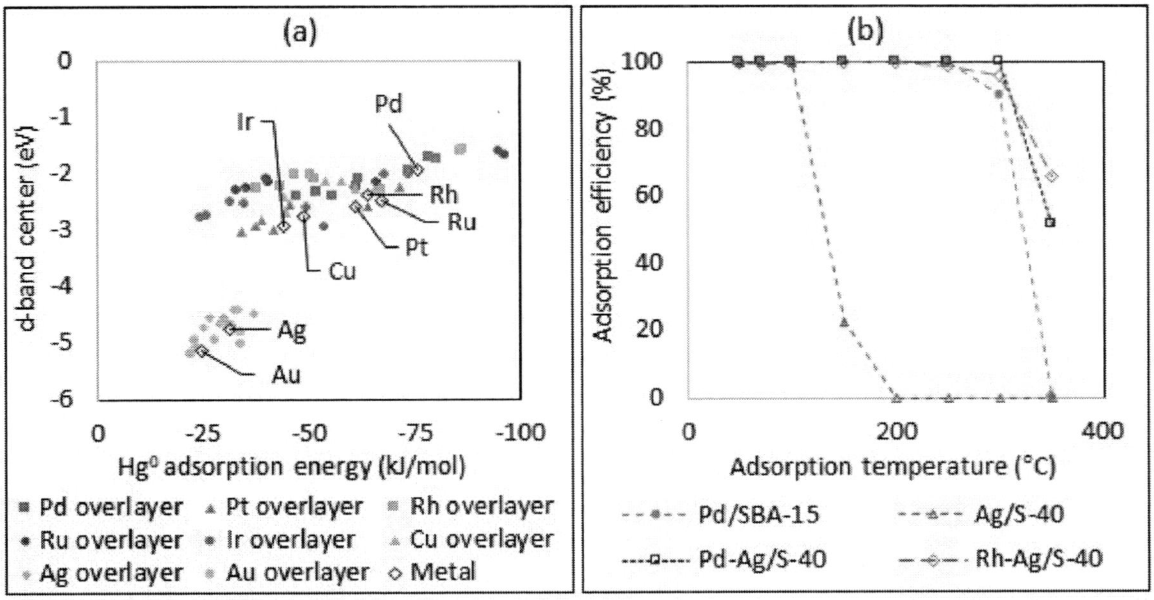

Figure 1. (a) Correlation between Hg^0 adsorption energy and d-band center of monometallic and overlayered bimetallic surfaces, (b) Effect of temperature on initial adsorption efficiency of lab-synthesized adsorbents.

9f Vapor-Liquid Equilibrium and Nucleation Studies of Water from First Principles-Based Machine Learning Models

Maria Carolina Nicola Barbosa Muniz, Roberto Car, Athanassios Panagiotopoulos

Throughout the years, much attention has been given to water in its bulk liquid and ice phases. The vapor-liquid equilibrium of water is involved in several relevant industrial applications, such as in power generation. When liquid water is superheated past the liquid-vapor coexistence line, it becomes metastable. The transition from liquid to vapor occurs subsequently with the emergence of small embryos (bubbles) of the vapor phase in a process termed bubble nucleation, and their collective collapse may lead to structural damage in industrial machinery[1]. There are significant challenges in studying these processes both experimentally and theoretically, as evidenced by the discrepancy of results obtained over the years in studies for properties such as nucleation rates and cavitation pressures[2]. Computer simulations can be employed to obtain a realistic description of these phenomena; however, results can be highly dependent on the chosen representation of the intermolecular interactions. In this work, we use a neural network energy function constructed with the Deep Potential Molecular Dynamics (DPMD)[3] framework, that accurately describes the SCAN approximation of quantum mechanical density functional theory. This model was recently used to predict the phase diagram of water's liquid and ice phases, obtaining a good overall agreement with experiment[4]. We apply this model to perform direct coexistence molecular dynamics simulations of water's vapor-liquid equilibrium and calculate properties such as coexistence densities, surface tensions and enthalpy of vaporization. Lastly, we describe ongoing efforts to perform a first principles bubble nucleation investigation of water to relate our results with previous studies.

References

1. Buckland, H. C., Masters, I., Orme, J. A. C. andamp; Baker, T. Cavitation inception and simulation in blade element momentum theory for modelling tidal stream turbines. *Proc. Inst. Mech. Eng. Part A J. Power Energy* **227**, 479–485 (2013).

2. Caupin, F. andamp; Herbert, E. Cavitation in water: a review. *Comptes Rendus Physique* **7**, 1000–1017 (2006).

3. Zhang, L., Han, J., Wang, H., Car, R. andamp; E, W. Deep Potential Molecular Dynamics: A Scalable Model with the Accuracy of Quantum Mechanics. *Phys. Rev. Lett.* **120**, 143001 (2018).

4. Zhang, L., Wang, H., Car, R. andamp; E, W. Phase Diagram of a Deep Potential Water Model. *Phys. Rev. Lett.* **126**, 236001 (2021).

9g Metastable Liquid-Liquid Criticality in Supercooled Wail Water

Jack Weis, Athanassios Panagiotopoulos, Pablo Debenedetti

The hypothesis that the anomalous behavior of liquid water is related to the existence of a second critical point in deeply supercooled states has long been the subject of intense debate. Recent, sophisticated experiments designed to observe the transformation between the two subcritical liquids on nano- and microsecond timescales, along with demanding numerical simulations based on classical (rigid) models parametrized to reproduce thermodynamic properties of water, have provided support to this hypothesis. A stronger numerical proof requires demonstrating that the critical point, which occurs at temperatures and pressures far from those at which the models were optimized, is robust with respect to model parameterization, specifically with respect to incorporating additional physical effects. Here we show that a liquid-liquid critical point can be rigorously located also in the WAIL model of water [J. Chem. Phys. 137, 014510 (2012)], a model parameterized using ab-initio calculations only. The model incorporates two features not present in many previously-studied water models: it is both flexible and polarizable, properties which can significantly influence the phase behavior of water. The observation of the critical point in a model in which the water-water interaction is estimated only using quantum ab-initio calculations confirms that the existence of two distinct liquids is a robust feature in the free energy landscape of supercooled water. Intriguingly, the WAIL model contains a critical point at milder temperatures and pressures than several previously-studied models.

Precise identification of this critical point is facilitated by several statistical-mechanical tools. First among these is the weighted-histogram analysis method, which allows observed distributions in thermodynamic properties – in this case, density and energy – to be reweighted to infer those at arbitrary temperatures and pressures. Those reweighted distributions are then compared to those observed in the 3D Ising model at its critical point: because both belong to the same universality class, the critical point can be identified when an appropriate linear combination of energy and density takes the same distribution as that of the magnetization of the critical Ising ferromagnet. This provides an objective function that can be optimized to find the critical temperature and pressure.

The existence of an LLT, and of the two distinct structures associated with it, also has implications for the nucleation of ice from the supercooled liquid. Nucleation from low-density liquid is substantially faster than from the high-density liquid, and the enhanced fluctuations associated with critical points are also known to speed nucleation. High supercooling enhances nucleation, but the approaching glass transition should frustrate it. This complex interplay gives rise to unexpected behaviors in the rate of ice nucleation near the critical point. The seeding technique was used, in combination with classical nucleation theory, to estimate these rates in the (less computationally intensive) TIP4P/Ice model of water.

9h Molecular Dynamics of Ethoxylated Surfactants in Water/N-Heptane Interface

Arthur M. Luz, Gabriel D. Barbosa, Thiago Jose Pinheiro Dos Santos, Carla L. M. Camargo, Frederico W. Tavares

Surfactants are substances widely used in several industrial segments. Mainly, their amphiphilic nature enables them to act as emulsifiers, drug carriers, detergents, and components for chemical flooding in enhanced oil recovery. Poly-(oxyethylene) alkyl ethers are a class of non-ionic surfactants often used as reference detergents due to their relatively simples structure compared to typical commercial surfactants. From the molecular point of view, their structure, referred to as $C_n(EO)_m$, is composed of a hydrophobic tail with n carbon units and a hydrophilic head consisting of m ethylene oxide units. This tunable structure allows a design to tailor the substance's properties to the desired application. To investigate the influence of structural changes, such as in the length of the hydrophobic tail and the number of ethoxylated units, we can conduct molecular dynamics simulations. With this technique, we can assess both macroscopic, such as interfacial tension and partition coefficients, and microscopic properties, such as density distributions, orientation profiles, and structural parameters, for water/oil interfaces in the presence of surfactants. Herein, we evaluate the interfacial properties of water/oil systems in the presence of poly-(oxyethylene) alkyl ethers surfactants through atomistic molecular dynamics simulations. We use the SPC/E model for water, and pure n-heptane represents the oil phase. The chosen surfactant structure was an hydrophobic tail composed of eight carbon atoms and an ethoxylated chain with a varying number of units. The GAFF force field was used for the oil phase and the surfactants molecules. We verify that the surfactant molecules exhibit tensoactive behavior, staying in the interfacial region and causing a reduction to the interfacial tension. As the number of ethoxylated units increases, so does the decrease in the tension. In terms of structure, we observe a folding behavior of the chains along with a parallel orientation in regards to the interface, maximizing their exposure to the interfacial region. Potential of Mean Force (PMF) calculations also show the relative solubility of these structures in both water and n-heptane phases. We observe that these surfactants are more soluble in n-heptane, while their solubilities in water increase with the size of the head group. We also verify that the free energy of adsorption to the interface in infinite dilution increases with the head group, as shown in the deeper free energy well calculated in the interfacial region.

9i Dynamic Molecular Switching for Environmentally Adaptive Surfaces

Nicholas Craven, Chris R. Iacovella, G. Kane Jennings, Clare McCabe

Monolayer films provide a coating through which surface interactions (e.g. water- and oil-repellent surfaces, lubricating films, biological implants, and adhesives) can be modified and tuned by changing the chemistry of the film.[1] Efforts have been made to design films that are responsive to changes in external medium;[2,3] however, such films are slow to respond or require external stimuli to change. Here we report molecular dynamics simulations of amorphous silica surfaces functionalized with short (6-18 carbons) alkylsilane chains terminated with functional groups that are composed of competing hydrophobic and hydrophilic subgroups. Such terminal groups can achieve a dynamic switching response through reorientation of the individual hydrophobic/hydrophilic subgroups which minimizes surface energy in response to contrasting environments. Changing the chemistry of the competing subgroups, chain surface density, chain length, and the presence of secondary chains (backfills) is examined to identify systems and conditions where dynamic switching occurs in response to changes in solvent. This vast parameter space is screened using the MoSDeF software suite[4] with the Signac[5] workflow manager.

References

[1] M. C. Stuart *et al.*, "Emerging applications of stimuli-responsive polymer materials," Nature Materials, **2010**, 9: 101-113.

[2] S. M. Kang, I. You, W. K. Cho, H. K. Shon, T. G. Lee, I. S. Choi, J. M. Karp, H. Lee, "One-Step Modification of Superhydrophobic Surfaces by a Mussel-Inspired Polymer Coating" Int Ed Engl. **2010**, 49(49): 9401–9404.

[3] Bo He and Junghoon Lee, 'Dynamic wettability switching by surface roughness effect,' The Sixteenth Annual International Conference on Micro Electro Mechanical Systems, **2003**, MEMS-03, pp. 120-123.

[4] A. Z. Summers, J. Gilmer, C. Iacovella, P. Cummings, C. McCabe, "MoSDeF, a Python Framework Enabling Large-Scale Computational Screening of Soft Matter: Application to Chemistry-Property Relationships in Lubricating Monolayer Films," JCTC., **2020**, 16: 1779-1793.

[5] C. S. Adorf, P. M. Dodd, V. Ramasubramani, and S. C. Glotzer. "Simple data and workflow management with the signac framework", Comput. Mater. Sci., **2018,** 146(C):220-229.

9j Interactions of Cellulosic Oligomers with Different Crystallographic Surfaces of Cellulose Nanocrystals through Molecular Simulation

Naveen Vasudevan, Dongyang Li, Li Xi

With added cellulosic polymers, cellulose nanocrystal (CNC) systems exhibit a plethora of enhanced properties due to non-trivial polymer-CNC surface interactions. Some of the properties include enhanced gelation, foam formation, etc. - In the presented work, we studied the interaction of two cellulosic oligomers namely, methylcellulose (MC) and cellulose or pentase (PE), with three crystallographic surfaces of CNC namely, (100), (110), and (110s) (where "s" indicates the surface is sulphated) in the presence of explicit water. Free energy of adsorption was calculated with umbrella sampling (US) combined with the COWBOE algorithm which we previously developed for optimizing the US parameter settings. The COWBOE algorithm was previously shown to give accurate potential of mean force (PMF) results with substantially lower computational time compared with the conventional approach. Extending this new method to wider polymer-interface combinations, as reported in this study, demonstrates its general applicability. We identified that the affinity of oligomers towards different surfaces ranks in the order of (100) > (110) > (110s). PE exhibited higher affinity towards (100) and (110s) surfaces compared with MC while both MC and PE displayed similar affinity towards the (110) surface. Detailed analysis of the oligomer-surface interactions, oligomer-solvent interactions, oligomer configuration and dynamics near the surface, and solvent distribution, etc., was performed to unveil the origin of the observed PMF differences in different oligomer-surface pairs. This investigation sheds light on the nature of the polymer-CNC interactions that are critical to the unusual property enhancement in those systems.

360a Exploring canyons in soft and glassy energy landscapes using metadynamics

Amruthesh Thirumalaiswamy, Robert Riggleman, John C. Crocker

The complex physics of glass forming systems is controlled by the structure of the low energy portions of their potential energy landscapes. The inability of glasses to equilibrate and find their lowest energy configurations is often attributed to their potential energy landscapes: rugged, barrier-filled surfaces in high-dimensional space that seem incredibly difficult to navigate. In this study, we report that a modified metadynamics algorithm (MIMSE) efficiently explores and samples low energy regions of such high-dimensional landscapes. In the energy landscape for a model foam, our algorithm finds and descends meandering 'canyons' in the landscape, which contain dense clusters of energy minima along their floors. Similar canyon structures in the energy landscapes of two model glass formers — hard sphere fluids and the Kob-Andersen glass - allow us to reach high densities and low energies, respectively. In the hard sphere system, fluid configurations are found to form continuous regions that cover the canyon floors up to densities well above the jamming transition. For the KobAndersen glass former, our technique samples low energy states with modest computational effort, with the lowest energies found approaching the predicted Kauzmann limit. In all three systems, our algorithm successfully uses these canyons as navigational aids to access low energy states deep within the potential energy landscape. The earlier-noted rugged surfaces line the canyon floors with smooth canyon walls; the algorithm successfully overcomes local ruggedness and explores low-energy regions of the landscape. Further, we observe that the underlying canyon have strikingly similar fractal character across these systems.

360c The Role of Heteroatom Placement and Chain Branching on Quaternary Ammonium Salts for Phase Transfer Catalysis

Brooks Rabideau, Rome Parker

Phase transfer catalysts (PTC) facilitate the movement of a reactant from one phase to another phase where the reaction occurs. These can reduce the need for volatile solvents such as DMSO or DMF for alkylation, cyanation, and oxidation reactions, and can also assist in the recovery of rare earth metals from electronic waste. Quaternary ammonium (quat) salts are frequently used because of their pervasiveness and the benefits that come with tuneability. Previous computational investigations on the relation between the structure and the action of quat salts as PTCs have focused on the role of alkyl chain length, anion strength, and the formation of water strands that develop during phase transfer. To date, there is still not a full molecular-level understanding of how heteroatom placement and chain branching effects the strength of ion pairing and the transfer between phases. Knowing how processes are affected by these structural properties will allow for the design of PTCs that target specific reactions and separations. In this work, molecular dynamics simulations (MD) are used in conjunction with umbrella sampling to determine the potential of mean force (PMF) associated with ion paring and the transfer between phases. Chain length, branching, and heteroatom placement are systematically varied and the PMF is determined. For several of the structures we also find the development of fine water filaments that develop as well as hysteresis that occurs during phase transfer. Further, we find that heteroatom placement and to a lesser extent branching have a significant effect on ion pairing and transfer, which are further discussed.

360d Investigating the Effect of Oxidic Functional Group on Graphene Oxide-Molybdenum Disulfide Heterostructures As Anode for Sodium-Ion Battery

Wonmyung Choi, Byungchan Han

The heterojuctions of layered materials are being massively studied as a potential electrode for a rechargeable sodium-ion battery (SIB). Among them, it is experimentally known that MoS_2/reduced Graphene Oxide (rGO) has higher capacity and lower electric conductivity than MoS_2/Graphene due to the functional groups. However, underlying mechanism for the changes in cell performances depending on functional groups has not yet been clearly identified. In this study, the effect of functional groups on capacity and metallicity is investigate using the first-principles density functional theory computation, which allows to decouple individual influence of a specific functional group. The results indicate that the Na intercalation energy becomes stable in the order of O, COOH, and OH, which implies elevated capacity. However, the electronic structure analysis shows opposite trend for, electrical conductivity. Our study provides guidelines for future development of high-performance electrodes through atomistic level understanding on the relationship between the functional groups in graphene and the functionality of electrode.

360e A Study on the Ionic Conductivity Improvement Mechanism of Halospinel Li2Sc2/3Cl4 By Variable Li and Dopant Concentration

SuSeong Hyun, Hoje Chun, Minjoon Hong, Byungchan Han

High performing solid-state electrolyte (SSE) with high ionic conductivity and electrochemical stability is of great interest for the next-generation lithium (Li) ion battery systems. Halide types have advantages with great ionic conductivity and relatively stable in operating voltage range compared to oxide and sulfide counterparts. Recently, one of spinel-type halide (Li2Sc2/3Cl4) has shown outstanding ionic conductivity of 2.07mS/cm. Here, we systematically investigate the aliovalent dopant (Fe^{2+}) effects on the Li-ion conductivity of Li2Sc2/3Cl4 to induce higher collective motion of Li by increasing Li-ion concentration. Also, we study the effect of Li-ion conductivities of different doping concentrations of Fe and Li in Li2Sc2/3Cl4. Stable configurations are obtained by density functional theory (DFT) calculations and underlying mechanisms of Li-ion diffusion are analyzed by ab-initio molecular dynamics (AIMD) simulations. The results reveal that the Fe-Li interaction plays a critical role in lowering the barrier of Li-ion motion along the diffusion path. Our design can guide to improve the performance with aliovalent dopings.

360f Redox Behavior of Single Atom Catalysts for the Upgrading of Plastic Waste-Derived Species

Jeremy Hu, Eun Mi Kim, Mikyung Hwang, Michael J. Janik, Hilal Ezgi Toraman, Konstantinos Alexopoulos

Within the last few decades, the production of plastics has exploded on a global scale. A vast majority of these plastics are disposed of in landfills or incinerated, while only 9% of the plastics ever made since 1950 have been recycled[1]. Efforts to redirect more plastic waste from landfill have led to a recent interest to find catalysts suitable to upgrade plastic waste-derived species to value-added products via selective hydrogenation reactions. Single atom catalysts, with active metals dispersed atomically rather than as extended clusters, are promising candidates due to their ability to hydrogenate specific functionalities to selectively form high value-added products. However, fundamental knowledge relating the electronic character of these systems with the energetics of selective hydrogenation elementary reactions remains lacking. Understanding the redox environment during catalytic upgrading reactions will ultimately guide the rational design of highly active and selective single atom catalysts.

Here, we use density functional theory (DFT) to elucidate how Cu or Ag single atoms on anatase TiO_2 can be selective for the hydrogenation of plastic waste-derived carboxylic acids. Specifically, we examine the electronic structure progression of benzoic acid redox processes on these catalysts. We study how the extent of reduction on the support will differentially stabilize species that aid in the selective formation of hydrogenation products. We also propose calibration schemes that quantify the extent of reduction along these elementary steps and how it ultimately relates to the selectivity of value-added products. Finally, a global optimization workflow is proposed that uses graphical representations of the surface and DFT to elucidate the morphology of these systems and how it directly affects redox behavior and catalyst performance.

1. Ritchie, H., and Roser, M. "Plastic Pollution" **2018**. *Published online at OurWorldInData.org*. Retrieved from: 'https://ourworldindata.org/plastic-pollution'.

360g Revisiting Nanoparticle Interactions for Effective Nanofluid Modeling and Simulation

Jee-Ching Wang

The interactions between nanoparticles are studied in this work with atomistic model surfaces and nanoparticles and with three atomistic potential models that were developed based on bulk properties. The results show that the Lennard-Jones potential of pair-wise nature tends to overestimate, while the Sutton-Chen and quantum Sutton-Chen potentials containing explicit many-body effects tend to underestimate. Regardless of the differences between the three potentials, the faceted nanoparticles are found to possess certain shape effects such that the interactions between the nanoparticles depend on the size of the surface facets. When characterized with the atomistically determined results, the Derjaguin approximation is found to be capable of providing interparticle interactions within the atomistically determined ranges and become a more reasonable approach for cases whose underlying interatomic interactions are of pair-wise nature. The Hamaker potential can also provide nanoparticle interactions with the right orders of magnitude. However, both of these classical coarse-grain approaches have the same tendency to underestimate nanoparticle interactions, especially at short distances where they overlook the significance of the discreteness of the interaction centers and excluded volumes. The same cause also makes them unable to exhibit the shape effects for the interactions between small faceted nanoparticles.

360h The Molecular Simulation Design Framework (MoSDeF): New Capabilities

Co D. Quach, Justin Gilmer, Umesh Timalsina, Nicholas C. Craven, Christopher Iacovella, Clare McCabe, Peter Cummings

The Molecular Simulation Design Framework (MoSDeF)[1] is a collection of python libraries designed to assist with the preparation of chemical, materials, and biomolecular systems for simulation using various open-source engines. The project was motivated by the irreproducibility crisis that exists in the science community, including the field of computational simulation. The software suite consists of three main components to reproducibly encapsulate different steps of the process of system preparation, namely, mBuild[2], Foyer[3] and GMSO (General Molecular Simulation Object).[4] mBuild[2] enables programmatic construction of system configurations using the concept of generative modeling that allows complex systems to be constructed from smaller, interchangeable pieces that can be connected together; mBuild allows for the creation of custom initialization scripts that can be shared and used as plugins. The Foyer[3] package performs atom-typing, i.e., the application of a force field to a given system; in Foyer, forcefield parameters are stored within XML files along with usage rules encoded as SMARTS strings, that are both human and machine readable, enabling the rules and parameters used to set up a simulation to be captured and disseminated. The GMSO[4] library is designed to store the fully parameterized system information, which can then be written out to syntactically correct input files for various simulation engines, including Monte Carlo engines (Cassandra, GOMC, and MCCCS) and molecular dynamics simulation engines (HOOMD-Blue, LAMMPS, GROMACS, and NAMD). The project has become a collaborative effort between multiple universities, where the collaborative development team has been working to add new features to these libraries, increasing their capabilities to support a wider range of chemical systems, configurations, simulation engines, and to ensure the interoperability with other existing and developing software packages. These efforts allow us to move closer to the goal of supporting Transferable, Reproducible, Usable by others, and Extensible (TRUE) simulation research.[5]

References

1. "MoSDeF" [Online]. Available: https://mosdef.org
2. Klein, J. Sallai, T. J. Jones, C. R. Iacovella, C. McCabe, and P. T. Cummings, "A Hierarchical, Component Based Approach to Screening Properties of Soft Matter", Foundations of Molecular Modeling and Simulation, **2016**, pp. 79-92.
3. Klein, A. Z. Summers, M. W. Thompson, J. B. Gilmer, C. McCabe, P. T. Cummings, J. Sallai and C. R. Iacovella, Formalizing atom-typing and the dissemination of force fields with foyer, Comput. Mater. Sci., 2019, 167, 215–227.
4. General Molecular Simulation Object (GMSO) [Online]. Available: https://github.com/mosdef-hub/gmso.
5. W. Thompson, J. B. Gilmer, R. A. Matsumoto, C. D. Quach, P. Shamaprasad, A. H. Yang, C. R. Iacovella, C. McCabe and P. T. Cummings, Towards molecular simulations that are transparent, reproducible, usable by others, and extensible (TRUE), Mol. Phys., 2020, 0, e1742938.

360i Microbubbles As Non-Invasive Targets for Blood Brain Barrier Disruption: An *in silico* investigation

Bailey Zinger, Payton Martinez, Kendra Kreienbrink, Mark A. Borden, Kayla Sprenger

Neurological disorders are the second leading cause of death globally; the four largest contributors are stroke, migraine, dementias such as Alzheimer's Disease, and meningitis [1]. In order to safely and effectively treat such neurological conditions, therapeutic agents targeting the central nervous system (CNS) must achieve passage through the blood-brain barrier (BBB). The BBB is a highly selective and semipermeable sheath of endothelial cells and tight junctions that regulates the movement of ions, molecules, and cells from the blood to the CNS [2]. Efflux transporters imbedded in the BBB also influence the rapid clearance of drugs from the CNS and impede drug localization in the brain [3]. The BBB is therefore a tremendous mechanical and chemical obstacle to the entry of therapeutic agents into the brain [4]. One promising method to transport drugs across the BBB and into the brain is to disrupt the structure of the BBB, temporarily and reversibly, via physical and chemical approaches. However, current methods of blood-brain barrier disruption (BBBD) and drug delivery to the CNS are evidenced with scarring, limited drug distribution, and potential mortality from cerebral edema [4]. Herein, we describe the use of molecular dynamics (MD) simulations to investigate the underlying mechanisms and utility of a novel and noninvasive approach to BBBD—microbubble (MB)-assisted focused-ultrasound (MB+FUS)—allowing for efficient, targeted delivery of drugs into the brain microenvironment.

MBs are one to ten microns in diameter and are composed of a lipid shell encapsulating a hydrophobic gas. MBs transiently permeabilize the BBB by cavitating under ultrasound, which leads to mechanofluidic intrusion of nearby tissue. Despite their promising development for BBBD, the exact underlying mechanisms of MB cavitation have yet to be understood. Recent experiments have shown that after undergoing compression via ultrasound shock waves, MBs experience rapid expansion from internal pressurized gas, leading to intriguing "patterning" behavior of MB surface lipids. We employed MD, coupled with enhanced sampling, to explore the origins of this behavior at the atomistic level. Specifically, we describe simulations of flat lipid monolayers—representing a magnified portion of the curved MB surface at the nanoscale—at a gas/solvent interface, and the application of an applied external force to mimic the effects of ultrasound-induced compression and expansion. Our results provide new insights into how MB surface lipids are able to hydrodynamically expand into their surrounding environment without destabilizing their shell structure, thereby minimizing damage to the BBB. These results thus demonstrate the utility of MBs for safely and effectively treating neurological disorders via BBBD.

[1] Daneman, R., andamp; Prat, A. "The blood-brain barrier", *Cold Spring Harb Perspect Biol*, **2015**, 7, a020412.

[2] Pardridge WM. "The blood-brain barrier: bottleneck in brain drug development", *NeuroRx,* **2005**, 2, 3–14.

[3] Löscher W and Potschka H. "Drug resistance in brain diseases and the role of drug efflux transporters", *Nat Rev Neurosci*, **2005**, 6, 591–602.

[4] Song KH, Harvey BK, and Borden MA. "State-of-the-art of microbubble-assisted blood-brain barrier disruption", *Theranostics*, **2018**, 8, 4393-4408.

360j Deep Reinforcement Learning As a Tool to Enable Coarse Grained Vaccine Models

Jonathan Faris, Daniel Orbidan, Brenden Petersen, Kayla Sprenger

Traditional approaches to vaccine development have yet to effectively compete with highly mutable infectious disease pathogens (hm-IDPs) such as HIV and influenza which present with rapidly evolving viral etiologies, which the human immune system cannot successfully manage alone. The limitations of these traditional approaches, paired with hm-IDPs ability to rapidly develop to bypass these efforts compound to one million deaths annually. To address the need for more effective vaccine development, computational agent-based models may be employed to simulate the complex interactions that occur between human immune cells and hm-IDP-like proteins (antigens) during affinity maturation—the process by which antibodies evolve. In contrast to existing experimental approaches, these computational models offer a safe, low-cost, and rapid means to study and assess the human immune response to vaccines, incorporating a wide range of design variables. However, the highly variable nature of affinity maturation and vast sequence space of hm-IDPs render standard agent-based modeling approaches alone insufficient for exploring all pertinent vaccine design variables and the subset of immunization protocols encompassed therein. To address this continuously diversifying sampling challenge, and to garner in tandem a better understanding of factors essential to vaccine development, we employed deep reinforcement learning to drive a recently developed coarse-grained agent-based model of affinity maturation to sample specific immunization protocols with the potential to improve the chosen metrics of protection (e.g., antibody neutralization potency or neutralizing antibody titers). With this approach, we were able to explore the *relevant space* of a wide range of vaccine design variables, giving rise to novel and testable insights into how vaccines should be developed to maximize protective immune responses to hm-IDPs in a manner that is more apt and flexible than traditional models. Furthermore, elucidating how these vaccine development models might be minimally tailored to account for major sources of heterogeneity in human immune responses.

360k Optiboost: A Method for Choosing a Safe and Efficient Boost for the Bond-Boost Method in Accelerated Molecular Dynamics (AMD) Simulations with Hyperdynamics (HD)

Jianming Cui

Accelerated molecular-dynamics (MD) simulations based on hyperdynamics (HD) can significantly improve the efficiency of MD simulations of condensed-phase systems that evolve via rare events. However, such simulations are not easy to apply since the optimal boosts are usually unknown. In this work, we developed a method called OptiBoost to adjust the value of ΔV_{max} (i.e., the boost) in HD simulations based on the bond-boost method. The OptiBoost method involves (1) Selecting a value of ΔV_{max} and running a series of short trajectories to obtain the probability $P(\Delta V_{max})$ that an event occurred; (2) Fitting $P(\Delta V_{max})$ to a sigmoid function and; (3) Delineating the boost range from the maximum in $P'(\Delta V_{max})$ and from the minimum in $P''(\Delta V_{max})$. We demonstrated the OptiBoost method in simulations on a cosine potential and applied it in three different systems involving Ag diffusion on Ag(100) in vacuum and in ethylene-glycol solvent. In all cases, OptiBoost was able to predict safe and effective values of the boost, indicating the OptiBoost protocol is an effective way to advance the applicability of HD simulations.

3601 Toward Accurate and Transferable Coarse-Grained Peptide Models Using Data-Driven Approaches

Luc Christians, Alexander J. Pak

So-called "bottom-up" coarse-grained (CG) modeling approaches attempt to retain essential microscopic information, e.g. as observed from atomistic simulations. Often, these methods are restricted to target system statistics at a given thermodynamic state point, such as temperature or pH, which limits the resulting model's transferability. For peptides and proteins with functions that vary with state point, such as peptide-based hydrogels, traditional CG modeling strategies would require extensive sampling and model derivations across a set of state points rendering such methods too computationally costly. In addition, current CG models that rely on pairwise interaction functional forms tend to lack fidelity in terms of conformational motion and/or long-range order. To address these problems, we use physics-informed data-driven approaches to derive implicit-solvent and low-resolution (e.g. one site per residue) CG models that retain accuracy across different state points; here, we focus on a set of alanine-rich peptides that exhibit conformational changes due to temperature or pH stimuli. We compare two methods to represent higher-order (i.e. beyond pairwise) CG interactions: (i) virtual sites that project solvent-mediated interactions [1] and (ii) deep learning neural network potentials, i.e. CGnet [2]. We present our analysis on CG models derived utilizing these methods across different state points. We discuss general strategies to represent state-point transferability with limited computational complexity, and describe how our approach could be extended to the greater family of natural amino acids. Ultimately, we envision that our CG models could be used to computationally investigate the structure, dynamics, and design of peptide systems, such as stimuli-responsive peptide-based hydrogels with applications in drug delivery or regenerative medicine.

Works Cited:

[1] A. J. Pak, T. Dannenhoffer-Lafage, J. J. Madsen, and G. A. Voth, "Systematic coarse-grained lipid force fields with semiexplicit solvation via virtual sites," *Journal of Chemical Theory and Computation*, vol. 15, no. 3, pp. 2087–2100, 2019.

[2] J. Wang, N. Charron, B. Husic, S. Olsson, F. Noé, and C. Clementi, "Multi-body effects in a coarse-grained protein force field," *The Journal of Chemical Physics*, vol. 154, no. 16, p. 164113, 2021.

360m Predicting Quantum-Accurate DNA Electron Densities and Forces with Equivariant Neural Networks

Alex Lee, Joshua A. Rackers, William P. Bricker

One of the fundamental limitations of accurately modeling biomolecules like DNA is the inability to perform quantum chemistry calculations on large molecular structures. We present a machine learning model based on an equivariant Euclidean Neural Network framework to obtain quantum-accurate electron densities for arbitrary DNA structures that are much too large for conventional quantum methods. The model is trained on representative B-DNA base pair steps that capture both base pairing and base stacking interactions. The model produces accurate electron densities for arbitrary B-DNA structures with typical errors of less than 1%. Crucially, the error does not increase with system size, which suggests that the model can extrapolate to large DNA structures with negligible loss of accuracy. The model also generalizes well to other DNA structural motifs such as the A- and Z-DNA forms, despite being trained on only B-DNA configurations. We show that this machine learning electron density model can be used to calculate electrostatic potentials of DNA with quantum accuracy. These electrostatic potentials produce more accurate results compared to classical force fields and do not show the usual deficiencies at short range. Lastly, the model can be used to directly calculate forces from the electron density through the Hellmann-Feynman theorem.

360n Elucidating Ethylene Hydrogenation on Site Isolated Intermetallic Catalysts

Angela Nguyen, Griffin A. Canning, Michael J. Janik, Robert Rioux

Selective hydrogenation of certain functionalities in molecules in mixed reactant streams are important in the chemical industry. By targeting specific functionalities unique to varying molecules, it is possible remove impurities in a stream without modifying the relevant reactants before continuing to further downstream processes. For example, selective hydrogenation of acetylene to ethylene can remove the acetylene impurity in ethylene prior to polymerization before the polymerization catalyst can be poisoned by acetylene. Currently in industry, Pd-based catalysts are used to catalyze the selective hydrogenation of acetylene, but the selectivity for ethylene is low.

Bimetallic catalysts are often used to improve selectivity performance as compared to single component catalysts. However, surface segregation of randomly distributed alloys is unavoidable in reaction conditions, which makes selective hydrogenation elusive. Intermetallics compounds can offer tunable site electronics and ensemble structure for selective hydrogenation catalysis. By tuning the composition of intermetallic compounds, it is possible to determine the site requirements needed for different functionalities. Pd-Zn γ-brass intermetallic compounds have been used to isolate the site requirements for ethylene hydrogenation varying from Pd-Zn-Pd to Pd-Pd-Pd by tuning the atomic composition. With these isolated active site ensembles, the ethylene hydrogenation mechanism on the Pd-Pd-Pd active site ensemble was fully enumerated allowing for a full detailed microkinetic modeled to be developed to predict reaction rates, surface coverages, and the degree of rate control. Herein, we expanded our work on Pd-Zn γ-brass intermetallic compounds to include a third coinage metal (M = Cu, Ag, Au) to further tune the active site ensemble to be (Pd-M-Pd), which has been shown to intermediate activity for ethylene hydrogenation when compared to the Pd-Zn γ-brass intermetallic compounds.

360o Electron Density Prediction with Graph Neural Networks on Large Catalyst Datasets

Ethan Sunshine, Muhammed Shuaibi, Zachary Ulissi

Large material datasets have recently been developed to build data-driven models for catalyst screening and discovery. Such datasets often contain a large swath of chemical systems. Graph neural networks (GNNs) have demonstrated their ability in accurately predicting molecular and material properties, notably energy and forces, but little work has been done to predict the electron density of these systems. Predicting the electron density is particularly challenging due to the large amount of data associated with a system in 3D space, O(100,000-1M+) grid points. In this work, we predict electron density using physically relevant descriptors which fit naturally into atom-centered GNNs. Namely, we fit a set of Gaussian multipole basis functions similar to those used in traditional computational chemistry calculations. This effectively reduces the electron density representation to a small number of node-level features in a graphical model. We also ensure this representation of electron density is equivariant to translations and rotations. These ideas are implemented as an extension to GemNet, a state of the art model for molecular and catalyst properties. We demonstrate our results on the Open Catalyst 2020 dataset (OC20), the largest catalyst dataset of its kind. Additionally, we show that including this additional physical information improves the overall performance of GemNet's energy and force predictions for OC20.

360p Exceptional Stability of Gd-Doped UO_2 Against Surface Oxidation: First-Principles Study

Minjoon Hong, Byungchan Han

An accurate understanding of structural integrity and chemical reactivity of UO_2 disposed in deep underground sites is of importance. Owing to the specific condition of the site location, UO_2 may have substantially different properties from the conventional prediction. In this study, we demonstrate that the oxidation resistivity of UO_2 is considerably modified by gadolinium (Gd), which is the element of neutron absorber and a byproduct of nuclear decay of radioactive U-235. Using density functional theory calculations, we investigate how the oxidation mechanism of UO_2 changes with Gd incorporation in U lattice. Our study indicates that Gd remarkably enhances the thermodynamic stability of pristine UO_2 against surface oxidation via three underlying mechanisms: (i) weakens the chemical bonding of adsorbed oxygen atom (O) with U, (ii) reduces active sites (U) for oxygen adsorption, and (iii) suppresses the subsurface diffusion of adsorbed O delaying the growth of the oxide layers on the UO_2. Electronic and lattice structure analyses for Gd-doped UO_2 indicate that amount of charge transfer from U to O is critically reduced and the lattice of the UO_2 surface is contracted. Our results provide useful information for understanding long-term stability and improving the structural integrity of UO_2 through the chemical doping process.

360q Explaining Improvements in Li-Ion Battery Performance By Atomic Layer Deposition of Alumina Using Molecular Dynamics Simulation

Julie A. Nguyen, Abigayle Becker, Krishan Kanhaiya, Alan Weimer, Hendrik Heinz

Atomic layer deposition (ALD) on layered cathode oxides like $LiCoO_2$ and $LiNi_{1-y-z}Mn_yCo_zO_2$ (NMC) has been shown to stabilize the surface and extend the cycling lifetime of these cathode active materials in lithium-ion batteries. While low cycle numbers of ALD have been experimentally shown to be beneficial to the lithium-ion cathode stability, the causes and mechanism for this stability are still elusive. We have utilized all-atom molecular dynamics simulations with accurate models for lithium Ni-Mn-Co oxides, as part of the INTERFACE force field (IFF) to study the alumina film morphology and the distribution of surface lithium through the film. We derive correlations between the number of alumina layers, the film crystallinity, the film thermodynamic stability, the lithium availability for charge transport at the film-cathode oxide interface, and the experimentally observed battery performance in terms of discharge capacity and number of ALD cycles .

Specifically, we studied alumina (Al_2O_3) deposition in contact with a layered cathode oxide (e.g., $LiCoO_2$ or other NMCs) without specific chemical modifications (such as hydration or carbonation products), and we modeled the trimethylaluminum (TMA)/water ALD chemistry with the cathode oxide surface after complete reaction. IFF in the 12-6 Lennard Jones form was used to model both the alumina and the layered cathode oxide. Using molecular dynamics simulations, we studied the structure, energy, and lithium distribution for zero to six monolayers of Al_2O_3: including multiple initial structures and annealing protocols to sample the configuration space and various degrees of simulated X-Ray crystallinity of the ALD films.

We find alumina films containing two to three monolayers exhibit favorable binding energies to the layered cathode oxide surface and optimum lithium availability on the exterior of the alumina film. Alumina layer numbers outside of two to three layers are thermodynamically less favorable. Generally, the more thermodynamically stable thin film-cathode interfaces possess lower alumina film crystallinities with the exception of two layers, which is highly crystalline, while thicker films are more crystalline. The lithium density profiles through the films show that the lithium exists on either side of thin Al_2O_3 films while the lithium becomes increasingly incorporated in the bulk of thicker films.

These atomic scale findings help explain experimental data showing two to four alumina ALD cycles yield a maximum discharge capacity and longer cycling stability compared to the uncoated NMC oxide surfaces.

360s Positive Unlabeled Learning of Peptide Properties

Mehrad Ansari, Andrew White

Deep learning can create accurate predictive models by exploiting existing large-scale experimental data, and guide the design of molecules. However, a major barrier is the requirement of both positive and negative examples in the classical supervised learning frameworks. Notably, most peptide databases come with missing information and low number of observations on negative examples, as such sequences are hard to obtain using high-throughput screening methods [1, 2]. To address this challenge, we solely exploit the limited known positive examples in a semi-supervised setting, and discover unknown peptide sequences that are likely to map to certain antimicrobial properties. In particular, positive unlabeled (PU) learning methods [3] effectively remove the dependence of deep learning classifiers on the negative training set, as they only requires a set of positive examples (confirmed sequence properties) along with an unlabeled set (the unknown candidate peptide sequences). We propose a two-step technique, where step 1 handles the deficiency of negative training examples by extracting a subset of the unlabeled data points that can be confidently labeled as negatives (i.e. reliable negatives). Subsequently, step 2 involves training a deep neural network classifier using the positive and the extracted reliable negatives, and applying it to the remaining pool of unlabeled examples. Our deep learning models share a common architecture of bidirectional recurrent neural networks to enhance the learning of bidirectional dependence between N-terminal and C-terminal amino acid residues [4]. We evaluate the predictive performance of our PU learning method and show that it can achieve competitive classification accuracy when compared with the state-of-art positive-negative (PN) classification approaches.

References:

[1] Song, Hyebin and Bremer, Bennett J and Hinds, Emily C and Raskutti, Garvesh and Romero, Philip A, "Inferring protein sequence-function relationships with large-scale positive-unlabeled learning", Cell Systems, 2021, 12, 92–101.

[2] Jowkar, Gholam-Hossein and Mansoori, Eghbal G, "Perceptron ensemble of graph based positive-unlabeled learning for disease gene identification", Computational biology and chemistry, 2016, 64, 263–270.

[3] Bekker, Jessa and Davis, Jesse, "Learning from positive and unlabeled data: A survey", Machine Learning, 2020, 109, 719–760.

[4] Ye, Yilin and Wang, Jian and Xu, Yunwan and Wang, Yi and Pan, Youdong and Song, Qi and Liu, Xing and Wan, Ji, "MATHLA: a robust framework for HLA-peptide binding prediction integrating bidirectional LSTM and multiple head attention mechanism", BMC bioinformatics, 2021, 1–12.

Positive-unlabeled learning: Some positive data are labeled, **all** negatives are unlabeled.

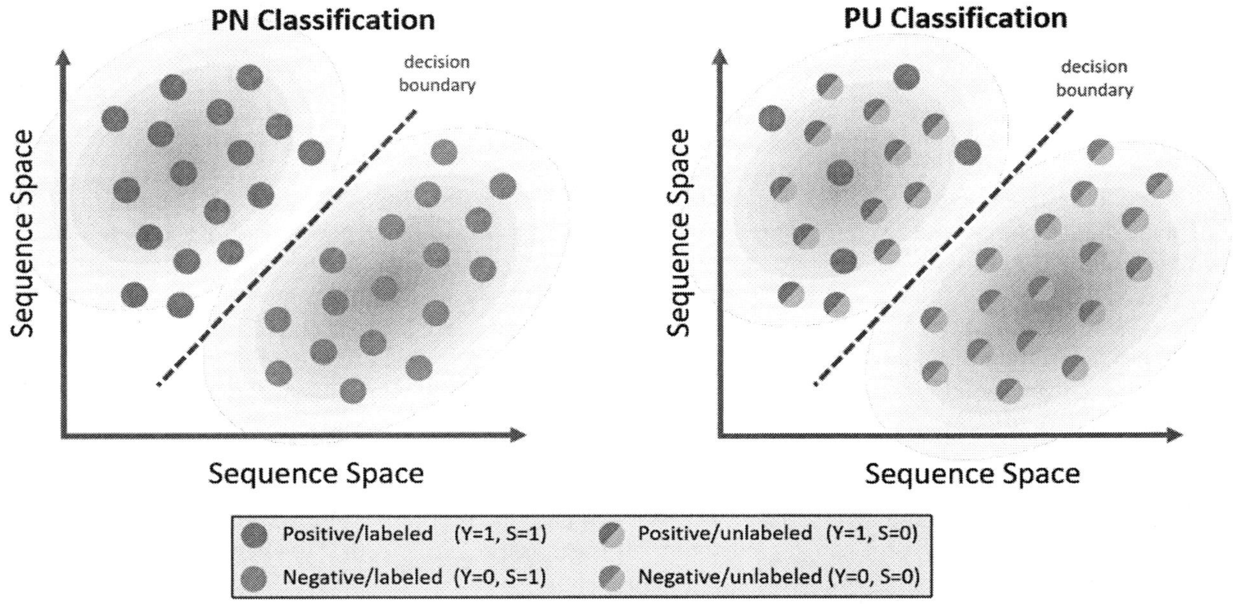

360t Elucidating Ligand Selectivity and Partial Agonism Towards Cannabinoid Receptors Using Machine Learning Approaches

Soumajit Dutta, Balaji Selvam, Austin Weigle, Diwakar Shukla

The endocannabinoid system (ECS) consists of the two cannabinoid receptors (CBs), their endogenous ligands, and enzymes. ECS maintains homeostasis in neuron signaling and modulates a plethora of physiological processes, including pain, obesity, inflammation, and cardiovascular regulation. Therefore, cannabinoid receptors are important targets for neurological and physiological disorders. However, drug development targeting cannabinoid receptors faces overstimulation and selectivity issues. Full agonists such as Fubinaca of the cannabinoid receptors cause dangerous side effects,including impairment of fine motor skills and increased blood pressure, tachycardia. Conversely, partial agonists such as Δ^9-tetrahydrocannabinol have been used as an appetite stimulant for AIDs patients and an antiemetic during chemotherapy with limited side effects. On the other hand, due to structural and sequence similarity, most of the drug molecules bind to both CB1 and CB2. Therefore, CB2 targeting molecules also bind to CB1, which is majorly expressed in the central nervous system (CNS) and causes psychological side effects. Thus, in this project, our objective is to develop a selective partial agonist using a data-centric machine learning approach. We build an extensive ligand activity dataset by curating information about ~6000 ligands from literature and databases. Using a graph neural network, we train our model to predict binding affinities of ligands for CB1 and CB2 separately. Our trained models show above 80% accuracy on the test dataset. These trained models were used to screen through molecules from the ZINC database to classify previously unknown molecules as selective CB-targeting partial agonists. Ultimately, our research will discover new scaffolds for developing selective partial agonists for therapeutic drug development.

360u Molecular Simulations of Protein/Ligand-Mediated Microglial Activation in Alzheimer's Disease

Emma Lietzke, Kimberley Bruce, Kayla Sprenger

Alzheimer's disease (AD) is a neurodegenerative disease that leads to dementia and is the 6th leading cause of death in the United States. A hallmark of AD is the buildup of amyloid-beta (A) plaques in the brain. Once activated, microglia, the key immune effector cells in the brain, seek out and cluster around these plaques, minimizing their neurotoxic effects. These "protective" functions are a result of ligand binding to microglial surface receptor proteins. However, due to the vast number of protein/ligand interactions that influence microglial activation, knowledge of which pathways result in protective microglial phenotypes remains elusive.

Triggering receptor expressed on myeloid cells 2 (TREM2) and lipoprotein lipase (LPL) are two microglial surface proteins that have been repeatedly implicated in the neuropathogenesis of AD. Herein, we describe the use of molecular docking and molecular dynamics (MD) simulations to investigate the interactions between TREM2 and LPL and determine how these interactions modify the cellular response to external stimuli (binding of different AD ligands). Our simulations illustrate novel potential binding mechanisms between TREM2, LPL, and other associated proteins in the cellular signaling pathway, which can then be confirmed experimentally through testing binding sites in HEK 293 and microglial BV-2 cells. Genetic variants of TREM2 related to increased risk for AD are also investigated to provide insights into how changes in these interactions mediate changes in microglial activation and phenotype via differential transmembrane signaling.

360v Coarse-Grained Molecular Dynamics Simulations of ssDNA Loaded Adeno-Associated Virus

Tibo Duran, Arani Chanda, Willow DiLuzio, Ryan Bellucci, Shivangi Naik, Bodhisattwa Chaudhuri

Purpose:

Adeno-associated virus (AAV) packaging a single-stranded (ssDNA) genome has emerged as a promising vector and is the most common vehicle for *in vivo* gene transfer. They are well established in clinical trials for *in vivo* gene therapy with a good safety profile, low immunogenicity, and long-term stability. However, little is known about the mechanism for the aggregation of AAV capsids during purification, their tendency to be absorbed by hydrophobic materials during the manufacturing process and storage, and the role of free ssDNA in the aggregation and adsorption processes. Therefore, we implemented coarse-grained molecular dynamics (CG-MD) simulations to probe the underlying mechanism of the interactions between capsids as well as the adsorption process of the capsid and hydrophobic surfaces, such as polyethylene (PE) and polypropylene (PP).

Methods:

The CG-MD simulations and their analyses were implemented using GROMACS software package with the MARTINI force-fields. The AAV Serotype 8 (AAV8) was chosen as the model viral capsid (PDB: 6v12) with 2.2kb ssDNA. The initial ssDNA model was constructed using the Avogadro software. The steepest descent algorithm was used in energy minimization followed by a 10 ns equilibration step in isothermal–isochoric ensemble at ambient temperature. The simulation production runs were performed in isothermal-isobaric ensemble using the Nosé–Hoover thermostat and the Parrinello–Rahman barostat at a pressure of 1 bar for each 20-fs time step. The simulations were carried out using periodic boundary conditions. Computations were performed in High Performance Center Supercomputer Cluster at the University of Connecticut.

Results:

The CG-MD simulations revealed the general behavior of capsid aggregation. The full capsid required more time to stabilize than the empty capsid under the same temperature and pressure conditions. This has been demonstrated by the analysis of the root-mean-square deviation (RMSD) and radius of gyration for the entire capsid, as well as root-mean-square fluctuation (RMSF) for the individual residues. Furthermore, a larger radius was captured for the full capsid than that of empty capsid, which is a resultant of the increased core size of loaded ssDNA molecule. In addition, due to the encapsulated negatively charged ssDNA, the full capsid has more anionic charges on the surface than that of the empty capsid, which was evaluated by the electrostatic potential map. Moreover, the behavior of a full capsid adsorbed onto the hydrophobic surface was captured successfully. It was noted that the capsid was rotated, driven toward, and attached to the surface with stronger interactions on the PP surface than that of the PE surface, which was assessed by the relative shape anisotropy.

Conclusions:

CG-MD simulations established a better understanding of capsid aggregation associated with multiple asparagine and glutamine residues. Furthermore, the adsorption mechanisms were revealed to be related to multiple hydrophobic residues, such as leucine, isoleucine and phenylalanine. By using this developed computational model, we were able to evaluate the stability and electrostatic charges of the capsids and free ssDNA molecule. Our simulations provide general dynamical information about the structure of the ssDNA and capsids on the microsecond timescale, revealing behavior that is impossible to be obtained from experimental studies. The CG-MD simulations are demonstrated to be a powerful tool in the simulation of macromolecular complexes and capable of guiding the interpretation, explanation, and direction of experiments.

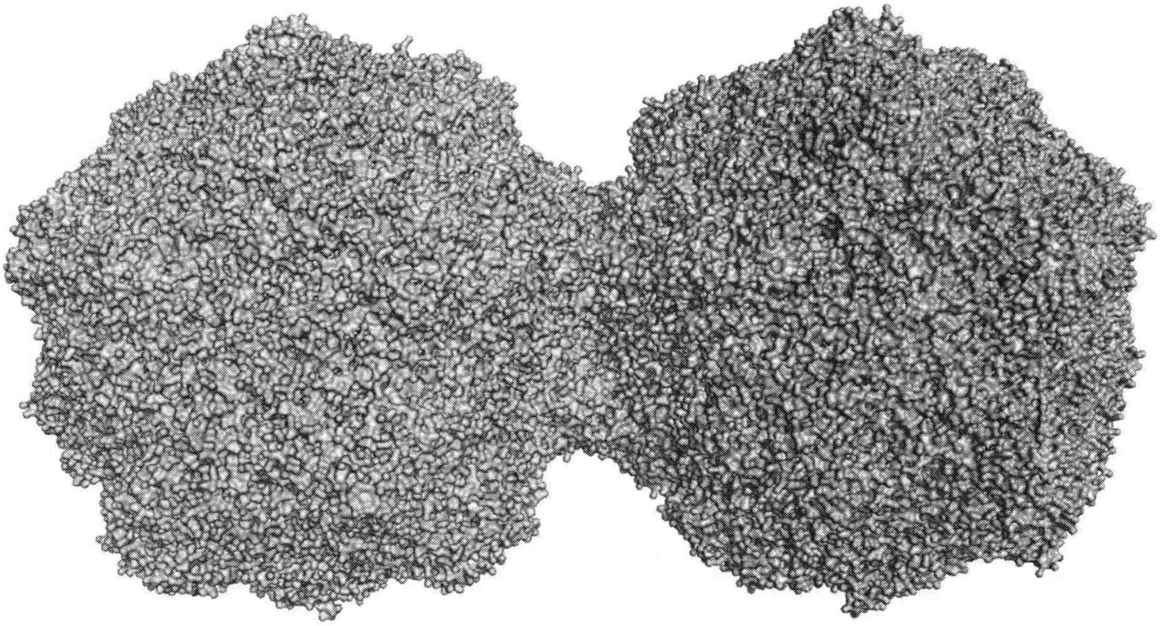

360w Design of Pore Wall Chemistry to Control Solute Transport and Selectivity

Sally Jiao, M. Scott Shell

Next-generation membranes for water treatment that move beyond simple desalination require materials with precisely tuned functionality. In particular, the purification and reuse of highly contaminated waters, such as oilfield-produced water, face key challenges in removing a wide array of solutes, including small neutral solutes that are difficult to separate. Multifunctional membrane surfaces potentially provide a vast, underexplored design space to improve membrane transport properties, but are difficult to design through trial-and-error. Here, we demonstrate an inverse design computational approach to efficiently identifying promising materials. We develop a combined optimization, machine learning, and molecular simulation workflow to engineer the transport of water relative to that of boric acid in a model nanopore by spatially patterning the pore wall with nonpolar methyl and polar hydroxyl groups. The genetic algorithm optimization identifies non-intuitive functionalization strategies that hinder the transport of boric acid through the pore, simply by altering the functional group patterning. Examining patterns inspired by the genetic algorithm results, we demonstrate that precise spatial functionalization of the methyl and hydroxyl groups differentially impacts the transport of water and boric acid in the pore, enabling design solutions that simultaneously improve both permeability and selectivity, a longstanding trade-off in membrane technology. This inverse design procedure and the accompanying molecular-level understanding of the effect of chemical heterogeneity on solute selectivity demonstrate new routes to the design of membrane materials with novel functionalities, and more broadly, show that surface chemical patterning can modulate water-mediated surface-solute interactions in systems relevant to many other technologies.

360x Studying Anhydrous Proton Transport on Graphene-Based Materials Using Deep Learning Methods

Siddarth Achar, Leonardo Bernasconi, Linfeng Zhang, Karl Johnson

The unique physical and chemical properties of graphene-based materials have spurred research interests in fuel cell applications, especially as proton exchange membranes (PEM). The design and study of such new proton conducting materials are essential for developing PEM fuel cells that operate at intermediate temperatures and conditions of low humidity. We show that graphane (hydrogenated graphene) functionalized with hydroxyl groups (graphanol) is a very promising material for anhydrous proton conduction. Previous methods to study proton conduction on graphanol involved performing computationally demanding density functional theory (DFT) calculations, limited to small systems and short time-scales. To overcome this computational hurdle, we developed dense neural network potentials using deep learning. Our previous work with using deep-learning potentials (DP) for graphane (the precursor material of graphanol) showed tremendous capabilities for capturing molecular physics with small training data sets. We have developed a DP for graphanol that is capable of accurately simulating both charged and uncharged systems. This DP predicts DFT forces to within meV/Å accuracy. We show that our DP is capable of accurately predicting vibrational and thermal properties, dynamic properties, and proton mobility of graphanol. We discuss active learning strategies required to achieve high accuracy for the graphanol DP.

Proton conductivity involves charge transfer and therefore cannot be fully described by observing motion of atoms alone. However, our graphanol DP potential does not include any information about electron density, charge polarization, or transport of charge centers. We have developed a deep learning formalism for predicting charge densities only using atomic positions for molecules and periodic systems. Our method, called DeepCDP (Deep-learning Charge Density Prediction), can predict electron densities for arbitrarily large systems. A DeepCDP model is trained on electron densities generated from DFT-MD simulations. Smooth Overlap of Atomic Orbitals (SOAP) is used to fingerprint atomic environments to their corresponding electron densities on a grid-point basis. Our DeepCDP model for graphanol achieves electron density prediction accuracies of >99% for both charged and uncharged systems. We demonstrate methods to use these predicted electron densities to compute the diffusion of the center of charge from DP simulations of graphanol with excess protons. Our combination of DeePMD and DeepCDP provides a complete description of proton conductivity *solely* using deep learning methods.

360y The Nuclear and Surface Electrostatic Potential As Descriptors of Chemical Interactions

Joakim Halldin Stenlid, PhD, Frank Abild-Pedersen

The electrostatic potential, $V(\mathbf{r})$, evaluated at the atomic nucleus or on the isodensity surface of a chemical compound can be used to make predictions about its local interaction tendencies. Chemical bonding is a central concept in chemistry and the ability to predict and explain interactions between chemical compounds allows the chemist to synthesize complex molecules and materials, estimate their chemical properties, and forecast their degradation. In this presentation, we discuss the advantages of using $V(\mathbf{r})$ for predictions of chemical interactions between adsorbates and material surfaces – interactions of importance in, e.g., catalysis, corrosion, and nanotechnology. $V(\mathbf{r})$ is a local property and its variations over a compound can be used to identify sites that are prone to interactions with either electron-acceptors (negative regions in $V(\mathbf{r})$) or with electron-donors (positive regions). The figure demonstrates the approach using a gold nanoparticle as example. Here, $V(\mathbf{r})$ is mapped on a 0.001 au isodensity surfaces – known as the surface electrostatic potential, $V_S(\mathbf{r})$. Red regions on the map are positive and found on atop sites. These interact with electron-donors such as the lone-pairs of H_2O, whereas negative (blue) areas found at hollow sites are susceptible to interactions with electron-acceptors such as the H-end of H_2O. Variations in the magnitude of $V_S(\mathbf{r})$ correctly ranks the interaction strength between the unique adsorption sites of the particle. In this presentation, we demonstrate that the analysis can be extended to a large variety of material particles, surfaces, and adsorbates. We will also show that the electrostatic potential at the nuclei of metal atoms on material surfaces is a strong indicator of the metal site stability. The stability, in its turn, is known to scale with adsorption energies of reaction intermediates in catalysis. Accordingly, it is found that the nuclear electrostatic potential is a good descriptor for chemical surface interactions.

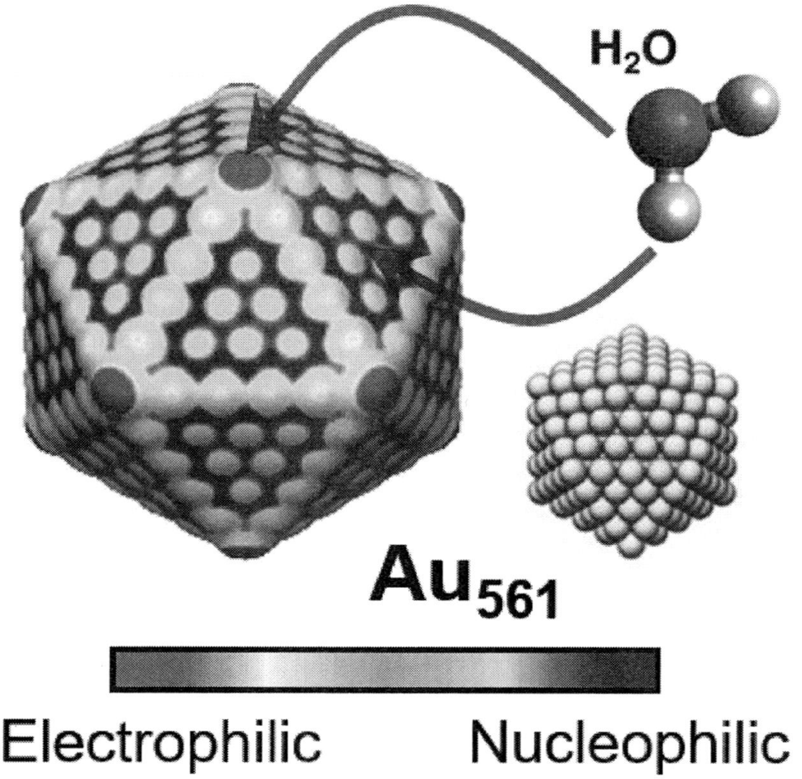

Figure. Surface electrostatic potential, $V_S(\mathbf{r})$, mapped on the 0.001 au iso-density contour. Based on density functional theory calculations.

360z Understanding DNA Hybridization through Thermodynamics and Kinetics of Abasic Oligomers

Mike Jones, Brennan Ashwood, Andrei Tokmakoff, Andrew Ferguson

The genetic and catalytic responsibilities of nucleic acids, as well as their more recent nanotechnology applications, rely on cooperative transitions of (un)folding and strand (de)hybridization. Nucleobase sequence plays a key role in all of these processes as it tunes the thermodynamic and time-dependent properties of nucleic acids, and a primary goal has been to develop a predictive understanding of sequence-dependent (de)hybridization. To interrogate this question, we performed computational and experimental analysis on a set of four DNA oligomers with a nucleobase removed from one of three sites along the duplex. This approach, analogous to residue mutations of proteins, allowed us to probe the role of individual nucleobases in the hybridization process as a function of sequence and relative location in the duplex. Using metadynamics and coarse-grained molecular dynamics, we converged a temperature-dependent free energy surface for each system, determined the relationship between the energy landscape and sequence and abasic location, and validated our finding using temperature-dependent IR spectroscopy and isothermal titration calorimetry. Furthermore, we constructed deep-learning optimized Markov state models and translated the collective system dynamics into interpretable mechanistic features with a distinct hierarchy of timescales. These features and timescales were compared against temperature-jump IR relaxation profiles and used to determine each oligomer's kinetic landscape and metastable state populations during (de)hybridization. Our combined experimental and simulation approach furnishes a detailed characterization of how an abasic site alters base pairing globally across small DNA oligonucleotide systems and provides insight into the contribution of individual nucleobases to DNA duplex hybridization. Furthermore, our findings can inform DNA nanotechnology design and provide insight into potential engineering applications for abasic DNA.

360aa The Chebyshev Interaction Model for Efficient Simulations (ChIMES): Machine-Learned Interatomic Models for Quantum-Accurate Reactive Simulation

Rebecca Lindsey, C. Huy Pham, Nir Goldman, Laurence E. Fried, Sorin Bastea

First principles methods such as density functional theory (DFT) are extremely powerful in their predictive power and ability to describe reactivity in atomistic simulations, however, high computational expense and poor scalability confines tractable problem space to time and length scales of a few hundred picoseconds and fewer than 10's of nanometers, respectively. While semi-empirical methods like density functional theory tight-binding (DFTB) can improve accessible timescales while retaining much of the predictive power of DFT, application of DFTB is still largely confined to sub-10 nm length scales. Thus, extensive work has gone into the development of reactive molecular mechanics force fields. For the vast majority of these models, however, the complicated functional forms required to accurately describe bond formation and breaking leads to challenging and time-consuming parameterization schemes, which, in turn, limits the physiochemical space over which parameters are available.

In this work, we present the Chebyshev Interaction Model for Efficient Simulations (ChIMES), a new reactive force field and fitting framework that retains most of the accuracy of DFT while decreasing computational requirements by several orders of magnitude. ChIMES models are comprised of *n*-body atomic interactions constructed from linear combinations of Chebyshev polynomials, and are entirely linear in fitting coefficients. Thus, model parameters can be rapidly generated by force matching to short DFT trajectories. Model details will be presented as will application to a variety of systems.

This work was performed under the auspices of the U.S. Department of Energy by Lawrence Livermore National Laboratory under Contract DE-AC52-07NA27344. Document release number LLNL-ABS-746816.

360ab Inverse Design of Open Nanocrystal Superlattices Using an Oscillating Pair Potential

Chase Petix, Michael Howard

Open nanocrystal superlattices are promising candidates for achieving materials with tunable electronic, optical, and mechanical properties, but there are limited scalable strategies for fabricating them. In this work, we computationally investigate the self-assembly of nanocrystals that interact through an oscillating pair potential mimicking long-ranged solvation forces between nanocrystals. We perform inverse design of the pair potential using relative-entropy minimization with our open-source inverse-design software package relentless, and we identify parameters that stabilize a variety of targeted lattice structures. This work may inform new experimental strategies for assembling tailored nanocrystal structures.

360ac Using Text-Mining and Community Knowledge to Quantify and Engineer Stability in MOFs

Aditya Nandy, Heather Kulik

For over two decades, metal-organic frameworks (MOFs) have been developed for various applications in gas separations, sensing, and catalysis. In practice, we must activate a MOF and remove solvent from its pores to render it porous and usable. Simultaneously, the MOF must also be stable under the thermal conditions. Although the tailored metal active sites and porous architectures of MOFs are promising for separations, sensing, and catalysis applications, a lack of understanding of how to improve their stability limits their use. MOFs vary in their coordination geometries, pore sizes, coordination chemistry, metal identity, and oxidation states, which challenge the development of general structure-activity relationships that generalize over various families of MOFs.

We use natural language processing (NLP) to extract stability information over thousands of MOF manuscripts, to obtain insights on solvent removal and thermal stability of these materials. Since computation cannot readily predict thermal or solvent removal stability, publicly available experimental data provides an avenue to construct machine-learning (ML) models. We use our graph-based representation, revised autocorrelations (RACs) for MOFs, to make predictions over a diverse set of MOFs. For the first time, we curate an extensive set of experimental data on MOFs that enable us to map MOF structures to their corresponding experimental stabilities.

We use our NLP derived labels on MOF stability to train ML models that predict MOF solvent-removal and thermal stabilities. From feature insights from our interpretable representation, we find that MOF stability is primarily governed by linker chemistry. Simultaneously, these feature analyses indicate the possibility of orthogonal tuning of stability, showing that solvent removal and thermal stability can be separately engineered. We show how hypothetical changes to MOF structures can be tested for their effects on stability using our ML models. We then demonstrate how we can use our ML models to propose alterations to 3*d* transition metal MOFs that can make them stable under catalytic conditions for methane oxidation. We also show how our strategy can lead to improved hyper-stable materials for MOF screening.

360ad Electrical Double Layer Capacitance and Polarizability Modeled Using Classical Molecular Dynamics.

Bolton Tran, Scott T. Milner, Michael J. Janik

Electrical double layers (EDL), which form in response to surface charge on an electrode in contact with electrolyte solution, influence the thermodynamics and kinetics of electrochemical processes. Atomistic modeling plays a key role in elucidating the physical characteristics of the EDL, which in turn can inform rational design and optimization for various important electrochemical systems such as electrocatalytic energy storage, capacitors, and fuel cells. In this poster presentation, we will present a classical molecular dynamics (MD) model for the EDL at the metal electrode and aqueous electrolyte interface. More specifically, we measure: 1) the double layer polarizability by calculating the dielectric constant and 2) the double layer capacitances at varying electrode potentials. In performing these measurements, we introduce two novel "add-ons" to the MD simulations: first is the description of the metal polarizability that we termed "QDyn," second is an alternating electric field (AC field) approach for measuring the dielectric response of water. For in-plane dielectric constant, we found the first water layer ($z < 0.5$nm) to exhibit a larger-than-bulk value ($\varepsilon \approx 110$). In contrast, the out-of-plane dielectric constant of this first layer is abnormally low ($\varepsilon \approx 3$). Next, we measure the double layer capacitance by varying the surface charge and assessing the corresponding electrostatic potential at the surface. Our MD-measured capacitance gives excellent agreement with the classical Gouy-Chapman-Stern (GCS) theory yet deviates from experimental values. We attribute this experimental deviation to the lack of charge spill-over from the electrode, which is a quantum effect that is missing from our classical model. Regardless, the dielectric constant and capacitance measurements made from our classical MD model are self-consistent through the GCS theory. Overall, our model provides a fully atomistic description of the EDL at adequate time and length scales, a feat that other state-of-the-art modeling tools still struggle with. As future work, this classical MD model will be combined with *ab initio* description of charge transfer processes to paint a more complete picture of the effect of EDL in electrochemistry.

360ae Insights of Phenolic Compounds Extraction from an Aqueous Environment Using Natural Deep Eutectic Solvents: Quantum Chemical and Molecular Dynamics Simulation

Nikhil Kumar, Tamal Banerjee

The phenolic compounds can be found in fruit, coffee, or tea and industrial wastewater from various industries and processes. The World Health Organization (WHO) has designated some of these phenolic compounds as priority pollutants because of their toxicity, so they must be removed before wastewater discharge. There has been a recent surge in interest in water-immiscible hydrophobic deep eutectic solvents (DESs). However, the hydrophobicity of these DESs has only been measured in a small number of studies. The stability of different DESs-water systems were studied to understand the stability of eutectic solvents to extract the phenolic compounds from aqueous environment. Based on relative stability screening, further molecular dynamics simulations studies of interfacial properties of DESs and water systems were performed, which plays critical role in application of liquid-liquid extraction(LLE). The considered DESs were DL-Menthol: Octanoic acid with a molar ratio of 1:1, DL-Menthol: Decanoic acid with a molar ratio of 1:1, Dodecanoic acid: Octanoic acid with a molar ratio of 1:3, and Dodecanoic acid: Decanoic acid with a molar ratio of 1:2. Thereafter, Non-bonded interactions, radial distribution function, combined distribution function, spatial distribution functions, and hydrogen bonding topology of various components were explored via MD simulations that highlighted the enhanced and favorable interactions of phenolic compounds (i.e., phenol and 2-chlorophenol) with hydrogen bond donor species of eutectic solvents. The charge-transfer (CT) process in quantum calculation confirmed the direction of CT from DESs to phenolic compounds, and the NBO analysis established the stability of DESs. The increase in bond distance between the HBA and HBD was confirmed with an increase in interaction between the DES and phenolic compounds. The atom-in-molecules (AIM) and noncovalent interaction (NCI) analysis suggest that strong noncovalent interactions are responsible for the higher extraction of phenolic compounds from water. The distribution coefficient (β), selectivity (S), and extraction efficiency (% EE) of phenolic compounds were calculated from the simulation and compared to the experimental LLE data. According to the MD simulation results, the extraction efficiency for the ternary DES-Water-Phenol or 2-chlorophenol systems was compared, which was in excellent agreement with experimental literature data with an RMSD of 1.7%.

360af Investigating the Electrocatalytic Reduction of 2,4,6-Tri-Nitro-Toluene (TNT) across Late Transition Metal Surfaces Using Density Functional Theory Methods

Andrew Wong, Joshua Miller, Brandon Perdue, Michael J. Janik

Remediation of 2,4,6-tri-nitro-toluene (TNT) to 2,4,6-tri-amino-toluene (TAT) is important in the application of environmental remediation, demilitarization efforts, and national security.[1] Electrocatalytic reduction of TNT is promising for these applications due to its potential to be an energy efficient process that operates at ambient reaction conditions. Recent experimental work investigating the reduction of TNT on Pd/Pt and glassy carbon electrodes (GCE) observed three distinct reduction peaks occurring at similar potentials across all electrodes, questioning the role of the electrocatalyst.[2] Lack of fundamental understanding of the elementary electrocatalytic mechanism inhibits the development of active TNT reduction electrocatalysts. We utilized DFT methods to determine the outer-sphere (non-catalyzed) and inner-sphere (catalyzed) elementary step mechanisms of TNT electrochemical reduction across late transition metal surfaces.

We observed that TNT reduction in the solution phase is inhibited by the initial reduction of the NO_2 group, where each subsequent reduction of the NO_2 group becomes more difficult. The non-catalyzed mechanism is potentially competitive for the reduction of the first NO_2 group when compared to the electrocatalytic reduction of TNT on the Fe (110) and Au (111) surfaces. Late transition metals for the electrocatalytic reduction of TNT are optimized by two activity tradeoff reactions: initial reduction of the adsorbed nitro species (NO_2-R*) and reduction of adsorbed hydroxyl species derived from TNT reduction (OH*). Using O affinity as the surface descriptor, we predict that late transition metal surfaces with intermediate binding of O*, such as Cu (111), and Pd (111), will exhibit higher activity towards TNT reduction. We extend our investigation to bimetallic and Fe_2O_3 (0001) surfaces as potential TNT reduction electrocatalysts.

References:

1. Ayoub, K., van Hullebusch, E. D., Cassir, M. andamp; Bermond, A. Application of advanced oxidation processes for TNT removal: A review. *Journal of Hazardous Materials* **178**, 10–28 (2010).
2. Soomro, R. A., Akyuz, O. P., Akin, H., Ozturk, R. andamp; Ibupoto, Z. H. Highly sensitive shape dependent electro-catalysis of TNT molecules using Pd and Pd–Pt alloy based nanostructures. *RSC Adv.* **6**, 44955–44962 (2016).

360ag Identification of Potential TMPRSS2 Inhibitors By Virtual Screening Using Molecular Docking and Machine Learning

Suraj Ugrani

Repeated viral outbreaks in the past and the recent COVID-19 pandemic have highlighted the urgent necessity for a broad-spectrum antiviral. Host proteases implicated in the infection mechanism of multiple viruses are ideal targets for such drugs. The human Transmembrane Protease Serine Type 2 (TMPRSS2), which is used by several coronaviruses including SARS-CoV-2, has emerged as a favorable target for treating coronaviruses. In this work, we perform virtual screening of small molecule protease inhibitors from the MEROPS database by molecular docking with UCSF DOCK to identify potential TMPRSS2 inhibitors. We also describe our homology model of the protease used for docking in the early days of the pandemic, which was found to be in close agreement with the subsequently resolved crystal structure. Compounds with the highest docking scores, which included approved anticoagulants like idraparinux and fondaparinux, were then analyzed for key interactions with TMPRSS2. In addition, we implement machine learning models, which use the results from docking to identify potentially important interactions and suggest modifications to improve the selected inhibitors. We continue to monitor emerging variants and new studies to ensure that our strategy remains pertinent.

360ah Self-Assembly of Lobed Colloidal Particles into Porous Morphologies

Brunno C. Rocha, Harish Vashisth

Colloidal particles, which are small solid particles suspended in fluid phases, can spontaneously reorganize themselves through a process known as self-assembly to form macrostructures with potential applications in the fields of materials science and biomedical engineering. The self-assembly of colloidal particles can be triggered by the addition of attractive patches on the surfaces of these particles. Moreover, control over the physical properties of the self-assembled macrostructures can be obtained through a careful examination of the number and locations of the patches on the surface of a particle. In this work, we performed Langevin Dynamics simulations of colloidal particles where the patches appear as solid protrusions (lobes), and the self-assembly is mediated by attractive interactions between the lobes. We studied the role of particle shape and long-range interlobe interactions in the formation of porous macrostructures across a range of temperatures. We also investigated the effect of polydispersity on the physical properties of the self-assembled structures. We observed that an intricate balance between the entropic and electrostatic interactions leads to the formation of macrostructures with distinct morphologies and porosities at different conditions. We also observed that the porosities of the self-assembled structures are enhanced when long-range electrostatic interactions are incorporated between the lobes. The evolving understanding on how to optimize the bottom-up design of lobed colloidal particles gained by leveraging simulation techniques can serve as a guide to further improve the manufacturing of porous scaffolds that better mimic biomaterials with desired properties.

360ai Simulation of Lipid Membranes Using Coarse-Grained Model and Reverse-Mapping

Hiroya Nitta, Taku Ozawa

Amphiphilic molecules such as lipids form a wide variety of membranes, including micelles, vesicles, lamellae, or more complex forms. The shape of the membrane depends on the temperature, concentration, and molecular structure of the constituent molecules. [1] Lipid membranes are studied using atomistic molecular dynamics (MD), coarse-grained molecular dynamics, dissipative particle dynamics, and simulations based on continuum models.

Atomistic molecular dynamics methods can provide detailed analytical methods at the atomic level. There are several tools to construct the initial structure of the membrane, such as CharmmGUI [2], Packmol [3],memgen [4], etc. However, often the problem is to know the equilibrium shape of the membrane under a given temperature, pressure and concentration. Moreover, the computational cost for membrane equilibration is very high. CG models can reach equilibrium with less computational effort than atomic models. Using accurate CG potential [5], the membrane structure can be obtained by relaxation simulation without any knowledge of final shape. However, the resolution of the CG model is not always sufficient. Therefore, one approach is to relax the membrane structure using a CG model and reverse-mapping to an atomistic model. This study describes a useful approach for membrane simulations using OCTA system [6] that includes initial particle location, relaxation MD simulation and reverse-mapping from CG to an atomistic model.

1. Arai et al., *Phys. Chem. Chem. Phys.*, 18, 19426 (2016)
2. Jo et al., *J. Comput. Chem.* 29, 1859-1865 (2008)
3. Martinez et al., *J. Comput. Chem.*, 30, 2157 (2009)
4. Christopher J. Knight and Jochen S. Hub, *Bioinformatics*, 31, 2897-2899 (2015)
5. Seo and W. Shinoda, SPICA, *J. Chem. Theory Comput.*, 15, 762 (2019)
6. OCTA, http://octa.jp

360aj Determination of Electronic Driving Factor for Selective Adsorption of Arsenic over Phosphorous Oxoanions By Fe(III)-Crosslinked Chitosan Using DFT

Obinna Nwokonkwo, Christopher Muhich

Adsorption-based water remediation is a promising "under-the-sink" technology for removing various contaminants from drinking water. Unfortunately, the co-occurrence of target contaminants and their less-toxic, iso-structural counterparts creates significant competition for the often-limited binding sites on conventional adsorbents, thus degrading removal efficiencies. In this work, we performed Density Functional Theory (DFT) calculations to understand the key factors that control the selectivity of arsenic oxoanions, $H_2AsO_4^-$ and H_3AsO_3, and phosphorous oxoanions, $H_2PO_4^-$ and H_3PO_3, by transition-metal crosslinked chitosan. We found that pure chitosan does not preferentially bind As(V) or P(V) since there was only a marginal difference of ~0.01eV in binding energies of both species. However, crosslinking chitosan with Fe (III) facilitates coordination-complex formation with the oxoanions. Analysis of the Crystal Orbital Hamilton Populations (COHP) revealed that the selectivity of As over P is controlled by the difference in the filling of antibonding molecular orbitals in the metal-ligand bond. Differences in localization of electron density in the HOMO resulted in stronger binding of the charged species, As(V) and P(V), over As(III) and P(III). These results show that slight variations in O partial charges of the oxoanions can provide a valuable tool to distinguish between nearly identical ions and, therefore, preferentially extract one over the other.

360an Development and Validation of Non-Bonded Interaction Parameters between Coarse-Grained Amino Acid Models and Water

Esmat Mohammadi, Soumil Joshi, Sanket Deshmukh

The interactions between amino acids and water are crucial in determining the behavior of many biological molecules as it may play an important role in protein folding, affecting their function. Thus, understanding these interactions at the molecular level is important for us to be able to control their function. Here, we present a set of non-bonded force-field (FF) parameters between 20 coarse-grained (CG) amino acid models and a one-site water model previously developed in our group. These non-bonded interactions between amino acid and water, which are represented by 12-6 Lennard-Jones potential, were obtained by employing particle swarm optimization (PSO) algorithm integrated with an artificial neural network (ANN). At first, we conducted all-atom (AA) MD simulations of the amino acid dipeptides, represented by CHARMM FF, in a TIP3P water box and calculated their Gibbs hydration free energies. Then, the non-bonded FF parameters between CG amino acid beads and the one-site water model were optimized to reproduce the hydration free energy of amino acids from AA MD simulations. The optimized parameters were then validated by utilizing them in CG simulations of the analogues of the amino acid side chains. The results show that the newly developed FF parameters could accurately capture the experimental hydration free energies of the analogues. This also indicates the transferability of the developed models for other molecules in CG MD simulations with water. Additionally, to further validate the accuracy of these parameters, we used the amino acid CG models in MD simulations of peptide amphiphiles to study their self-assembly behavior as well as to study polypeptides in solvents.

360ak Insights into the Phase Diagram of Pluronic L64 Using Coarse-Grained Molecular Dynamics Simulations

Mangesh Bhendale, Arpita Srivastava, Jayant K Singh

We investigate the concentration dependent phase diagram of pluronic L64 in aqueous media at 300 K and 320 K using coarse grained (CG) molecular dynamics (MD) simulations. The CG model is derived by adapting the Martini model for non-bonded interactions along with the Boltzmann Inversion (BI) of bonded interactions from all atom (AA) simulations. Our derived CG model successfully captures the experimentally observed micellar, hexagonal and 2D phases. The end-to-end distance reveals the configurational transition from open-chain to folded-chain structure from micellar phase to lamellar phase, increasing the orientational order. Increase in temperature leads to expulsion of water molecules from L64 moiety suggesting increase in L64 hydrophobicity. The thermodynamic analysis using two phase thermodynamics (2PT) method suggests the entropy of the system decreases with increase in L64 concentration and the decrease in free energy (F) with temperature is mainly driven by entropic factor (-TS). Further, the increase in aggregation number at lower concentration and self-assembly at very high concentration is energetically driven whereas the micellar to lamellar transition is entropically driven. Our model provides the molecular insights into L64 phase transitions which can be further explored to design functionality based supra-structures for drug delivery and tissue engineering applications.

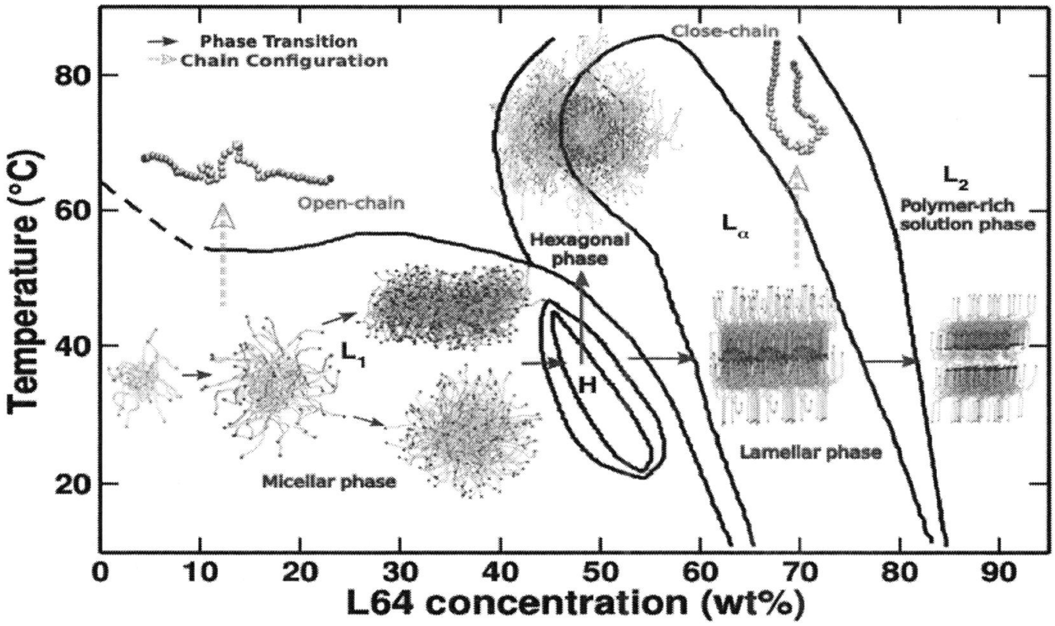

360ao Elucidating the Bulk and Interfacial Structure of Ionic Liquids from the Dilute to Concentrated Regimes Using Molecular Dynamic Simulations

Lisa Je, Beichen Liu, Matthew Gebbie, Victor Zavala, Reid Van Lehn

Ionic liquids (ILs), such as 1-ethyl-3-methylimidazolium tetrafluoroborate (EMIM-BF4), have been scrutinized as viable electrolytes for electrochemical reactions due to their highly tunable cation/anion properties, intrinsic conductivity, wide electrochemical stability window, low volatility, and strong electrostatic interactions. While these features make ILs promising, the molecular-scale structure of ILs in bulk solution and near electrochemical interfaces remains poorly understood, inhibiting the rational application of ILs for electrochemical reactions. To address this gap, molecular simulations can be utilized to study the dynamic nature of IL structure that is difficult to observe in experiments, especially as a function of concentration and near the electrode surface. In this work, we performed atomistic molecular dynamics (MD) simulations to quantify concentration-dependent changes in EMIM-BF4 structure in solvents including acetonitrile, dimethyl sulfoxide, and propylene carbonate and determine how interactions between the IL and solvent impact structure. We show that EMIM-BF4 exhibits variations in the tendency to form molecular-scale clusters as a function of concentration ranging from dilute to neat regimes. We further parameterized descriptors of ILs for physiochemical property prediction related to IL structural clustering and report variations in structural parameters induced by the presence of an interface. These quantitative metrics will permit analysis and understanding of how IL structure impacts electrochemical reactions, which is necessary for fine-tuning electrolyte properties for these systems.

360ap Developing Deep Learning Models to Predict Sigma Profiles of Lignin-Derived Organic Molecules

Usman Abbas, Manh Tien Nguyen, Yuxuan Zhang, Jian Shi, Jin Chen, Qing Shao

Lignin-derived molecules have been considered as a rich pool to search for suitable solvents or other functional materials. One unmet need is to evaluate which molecules may possess the desired properties. Sigma profile is the numerical integration of the electrostatic potential map of a given molecule. The sigma profile has been recognized as an important feature to evaluate the potential function of an organic molecule. However, the calculation of sigma profiles usually relies on time-consuming quantum mechanical calculations, limiting its application. Machine learning methods can bypass the time-consuming quantum mechanical calculations by directly correlating the representation of a molecule with its sigma profile. In this work, we develop machine learning methods to predict the sigma profile of lignin-derived organic molecules based on their SMILES strings. We first trained an encoder-decoder model on the unlabeled SMILES representations of molecules from CHEMBL. The encoder-decoder model converts the SMILES string to the vectors on the latent molecular space. Then we develop a neural network to predict sigma profiles of molecules based on the vectors on the continuous molecular space using the VT-2005 database. At last, we evaluate the performance of the developed model based on its ability to predict the sigma profile of a group of lignin-derived organic molecules.

360aq Effect of Solvent Quality on Structure and Dynamics of Lignin in Solution

Nusrat Jahan, Md Masrul Huda, Neeraj Rai

Due to significant aromatic content, lignin is an attractive source of valuable organic chemicals. As the majority the proposed lignin depolymerization are expected to be liquid-phase, it is imperative to understand the effect of solvent quality on the structure and dynamics of lignin. Here we use all-atom molecular dynamics simulations to understand the evolution of lignin structure as a function of concentration of methanol in methanol/water solution at different temperatures. We utilize two different lignin models, softwood that consists of guaiacyl (G) monomer, and hardwood that consists of heteropolymer containing guaiacyl:syringyl (S) with a 1.35:1 ratio. The presence of methoxy groups in the hardwood leads to a slightly higher value in the Flory exponent with increasing methanol concentration. Furthermore, softwood lignin adopts the ideal chain state in 80 mole% methanol solution while the hardwood adopts the same state in 40 mole% mixture at 300 K. We find that methanol preferentially solvates the nonpolar segments of the lignin polymer while water molecules solvate the polar functional groups. Thus, as the methanol concentration increases, methanol is able to better solvate lignin polymer leading to more extended configuration.

360ar Identifying the Stoichiometry of the Metastable Cu^{3+} State in Alkaline Electrochemical Systems Using DFT-Based Theoretical Raman Standards

Lars Ostervold III, Bolton Tran, Maxwell Wetherington, Konstantinos Alexopoulos, Lauren F. Greenlee, Michael J. Janik

While normally used for CO_2 reduction, copper electrocatalysts have recently shown promise for biomass upgrading[1] and the oxygen evolution reaction (OER), with overpotentials comparable to the state-of-the-art OER catalysts[2] at 10 mA-cm^{-2}. However, likely due to the comparative lack in popularity for copper OER catalysts, the active site for the OER on copper is unknown, hindering further fundamental studies. The demonstrated low overpotentials for CuO-based materials combined with the low cost and high availability of copper justify further investigation into copper OER catalysts. A study by Deng et al.[3] linked a distinct Raman signal to the active site for copper-based OER but did not provide further insight into the chemical nature of the active site. The metastability of this species makes investigations difficult, but here we solve the problem by using theoretical Raman standards based on density functional theory (DFT). First, we performed operando Raman during linear sweep voltammetry (LSV) to note a unique peak corresponding to a formal Cu^{3+} species ca. 587 cm^{-1}. We then screened for possible components by simulating the Raman spectra of ~30 structures with formal copper charges ≥2+. We further narrowed the possibilities by performing additional operando Raman experiments along with a series of cyclic voltammetry studies - providing a compelling case for a CuOOH species.

After identifying the CuOOH species, we performed DFT calculations to determine the electronic structure of this CuOOH species as d^9L Cu^{2+} (where d^n denotes the n number of electrons in the d orbital of copper and L denotes an electron hole in the ligand, oxygen) by Bader charge and density of state analysis. This is similar to other formal Cu^{3+} species, where electron density is removed from the bonding p-orbital due to inverted ligand behavior.[4] We also performed calculations of hydroxide adsorption on a CuO(111) facet and provided an explanation for the background oxidation current between the Cu^{2+} and formal Cu^{3+} states. There are two unique copper atoms in the experimentally-verified CuO crystal structure with one of these unique atoms being more accessible to solution species. The hydroxide species prefer adsorption on the bridge site between the surface-exposed copper atoms (E_{ads} ~0.62 eV). However, there is a surface-coverage dependency that will increase the E_{ads} up to 1.92 eV for total site coverage. We suggest the background oxidation current between the two formal oxidation states is due to continuous partial oxidation of Cu atoms by hydroxide adsorption, which requires increased applied potentials as surface coverage increases. We hope the identification of the metastable Cu^{3+} species and other atomic insight leads to further rational design of copper electrocatalysts.

(1) Ostervold, L.; Perez Bakovic, S. I.; Hestekin, J.; Greenlee, L. F. Electrochemical biomass upgrading: degradation of glucose to lactic acid on a copper(ii) electrode. *RSC Advances* **2021**, *11* (50), 31208-31218, 10.1039/D1RA06737K. DOI: 10.1039/D1RA06737K.

(2) Roger, I.; Shipman, M. A.; Symes, M. D. Earth-abundant catalysts for electrochemical and photoelectrochemical water splitting. *Nature Reviews Chemistry* **2017**, *1* (1). DOI: 10.1038/s41570-016-0003.

(3) Deng, Y.; Handoko, A. D.; Du, Y.; Xi, S.; Yeo, B. S. In situ Raman spectroscopy of copper and copper oxide surfaces during electrochemical oxygen evolution reaction: identification of CuIII oxides as catalytically active species. *Acs Catalysis* **2016**, *6* (4), 2473-2481.

(4) Steen, J. S.; Knizia, G.; Klein, J. sigma-Noninnocence: Masked Phenyl-Cation Transfer at Formal Ni-IV. *Angewandte Chemie-International Edition* **2019**, *58* (37), 13133-13139. DOI: 10.1002/anie.201906658. DiMucci, I. M.; Lukens, J. T.; Chatterjee, S.; Carsch, K. M.; Titus, C. J.; Lee, S. J.; Nordlund, D.; Betley, T. A.; MacMillan, S. N.; Lancaster, K. M. The myth of d8 copper (III). *Journal of the American Chemical Society* **2019**, *141* (46), 18508-18520.

360as Bayesian Forcefield Driven Monte Carlo and Molecular Dynamics Simulations of O and Cl Promoted Silver Surface Reconstructions

Anna Sviripa, Ching-Tien Chen, David Flaherty, Christopher Paolucci

Silver nanoparticles (NPs) are commercial catalysts used for selective oxidation of ethylene to ethylene oxide (EO). Higher selectivity towards EO formation is achieved by incorporation of chlorine promoter. Ultra-high vacuum studies indicate that the Ag surface undergoes significant reconstruction upon exposure to O_2, or mixtures of O_2 and Cl. However, the surface structure formed by contact between Ag NPs and O_2 at relevant EO conditions remains debated, in both the presence and absence of Cl. Atomistic modeling of surface reconstructions with density functional theory (DFT), is challenging due to spatial and temporal limitations associated with electronic structure calculations. Here, we use DFT sampling of potential energy surfaces implemented in the Fast Learning of Atomistic Rare Events (FLARE) code [1] to generate a Gaussian Process (GP) interatomic potential descriptive of O and Cl promoted Ag surfaces.

To sample a diverse distribution of Ag-O and Ag-O-Cl interactions we performed: 1) on the fly active learning with *ab initio* molecular dynamics (MD) simulations of O and Cl single and mixed coverages on Ag(111) and $Ag_2O(001)$ slabs, in the NPT ensemble, at 1500 K for 5 ps; 2) offline learning of DFT optimized reconstructed Ag surfaces proposed in literature. Trained models were validated with DFT calculations of structures sampled from the GP driven MD and Monte Carlo (MC) simulations, which were not included in the training data. The Ag-O trained model resulted in a RMSE of ~10 meV/atom for energy and 0.2 eV/Angstrom for forces comparable to machine learning forcefield errors reported in the literature.

To determine how Ag surfaces dynamically evolve under different conditions, we performed GP driven NVT MD simulations on the Ag(111), (110), (100) and reconstructed c(4x8) surfaces at different temperatures and O coverages in LAMMPS. We observed significant surface reconstruction and O subsurface diffusion for all surfaces. We analyzed all simulated surfaces by computing the time-averaged radial distribution functions and found that regardless of the initial state, all reconstructions reach similar final environments. Next, to overcome the activation barriers inherent to MD, we performed NVT MC simulations using an inhouse code. The code implements swap moves, which allow for subsurface diffusion, followed by geometry optimization, which minimizes atomic forces and increases probability of moves being accepted. In NVT MC simulations we observed surface reconstruction, O subsurface diffusion and formation of surface dioxygen species. The dioxygen species formation on Ag surfaces containing subsurface O, which we observed in MC simulations only, is consistent with recently reported experiments and DFT calculations [2,3]. We also performed grand canonical ensemble MC to simulate equilibrium surface coverages of atomic O and dioxygen species at varying temperatures and chemical potentials of oxygen, and related these to oxygen pressures used experimentally by our collaborators. Finally, for the Ag-O-Cl system we trained a GP forcefield and used it for both MD and MC. Our preliminary results suggest that the Ag surface reconstructs and both O and Cl species diffuse subsurface. This work improves our understanding of how both separately and together O and Cl affect Ag surface dynamics under EO reaction conditions.

References

1. Vandermause, J., Torrisi, S. B., Batzner, S., Xie, Y., Sun, L., Kolpak, A. M., Kozinsky, B,. npj Comput Mater 6, 20 (2020).

2. Liu, C., Wijewardena, D. P., Sviripa, A., Sampath, A., Flaherty, D. W., Paolucci, C., J. Catal, 405 (2022).

3. Pu, T., Setiawan, A., Mosevitzky Lis, B., Zhu, M., Ford, M. E., Rangarajan, S., Wachs, I. E., ACS Catal., 12 (2022).

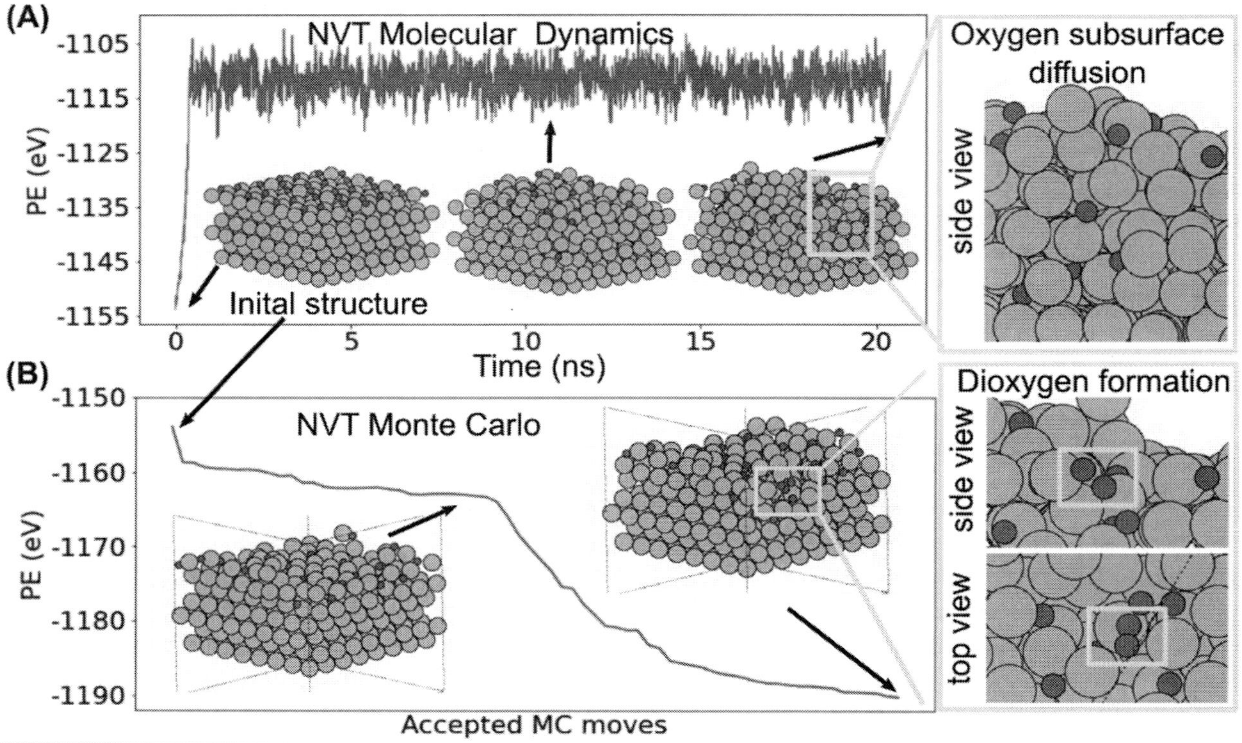

Figures: A) Potential energy vs. time for GP driven NVT MD simulation; B) Potential energy vs. accepted MC moves for GP driven NVT MC simulation.

360at Explosive Mechanochemistry: Foundations for Strength-Aware Chemical Kinetics

Matthew Kroonblawd, Brad Steele, Brenden Hamilton, Chunyu Li, Matthew Nelms, Ryan Austin, Laurence E. Fried

Molecular dynamics shows strong synergy between high-rate strength behavior and chemistry in explosives, with plastic deformation localizing heat into hot spots and reducing reaction barriers through mechanochemistry. Nanoscale shear bands form dynamically in many shocked explosives, and we show that these regions are 200 times more reactive than bulk crystal in TATB (1,3,5-triamino-2,4,6-trinitrobenzene). Enhanced reactivity is traced to molecular deformations that are identified based on intramolecular strain energy. Axial compression simulations show that substantial molecular deformations are generated for all crystal orientations at pressures greater than 10 GPa, which qualitatively explains unusual changes in TATB reactivity with increasing shock pressure. Despite fine-scale complexity, the global stress-strain response is straightforward with a flow stress that scales with the shear modulus. This suggests a plausible route to simplify treatments of strength, which when coupled to our measured intramolecular strain energy distributions, provides grounds for parameterizing a "strength-aware" chemical kinetics model for TATB-based explosives.

This work was performed under the auspices of the U.S. Department of Energy by Lawrence Livermore National Laboratory under Contract DE-AC52-07NA27344. Approved for unlimited release, LLNL-ABS-828054.

360au Protein Interactions Determined from Computational Methods

Gregory Dignon, Barbara Hribar-Lee, Dima Kozakov, Ken A. Dill

Biomolecular interactions are essential for biology, and incredibly prevalent in biotechnology. Similarly, biomolecular interactions can result in problematic behaviors, both in biology and biotechnology. Thus an understanding of how shape, chemistry and complementarity between two biomolecules dictates their interactions yields huge benefits in many fields such as structure-based drug design. Current computataional methods have seen huge success in predicting protein structures as well as protein-protein interaction orientations, while it is much more cumbersome to determine binding affinities between two large biomolecules. Using coarse-grained theories and simulation techniques in conjunction with fully atomistic simulations, we present an approach to calculate the relative binding affinities of biomolecules in the presence of various solution additives.

360av A Computational Study for Predicting Stability Differences in Multiple Conformations of the Sars-Cov-2 Frameshifting RNA Element

Karim Malekzadeh, Gul H. Zerze

Pseudoknotted RNA structural elements stimulate the frameshifting process during protein translation. The RNA pseudoknot that causes ribosomal frameshifting in SARS-CoV-2 is a potential drug target for fighting against Covid-19. This small part of RNA programs translational frameshifting during the translation of RNA from Open Reading Frame (ORF) 1a upstream to 1b downstream. Viral polyproteins which are produced in this process are essential in viral replication and any interference in this process, including structural modifications, will alter virus replication rates. Recent advances
in this area for predicting two-dimensional (2D) structures have shown that the frameshifting element can take multiple conformations [1]. Despite this progress in structure prediction, we still lack an important piece of information: stability differences between 3D structures of this molecule. In an effort to address that computationally, in this work, we have adapted our recently developed technique [2] to calculate the free energy surface of 13469−13545(77 nt) of SARS-CoV-2 RNA to measure the free-energetic competition between 3-stem H-type pseudoknot (3_6) and HL-type 3-stem pseudoknot (3_3) by using a combined parallel tempering and metadynamics technique.

[1] Zerze, Gül H., Pablo M. Piaggi, and Pablo G. Debenedetti. “A Computational Study of RNA Tetraloop Thermodynamics, Including Misfolded States.” The Journal of Physical Chemistry B 125.50 (2021): 13685-13695.

[2] Schlick, Tamar, et al. 'To knot or not to knot: Multiple conformations of the SARS-CoV-2 frameshifting RNA element.' Journal of the American Chemical Society 143.30 (2021): 11404-11422.

360aw Comparison of Advanced Sampling Techniques for Atomistic Scale RNA Folding

Kosar Rahimi, Gul H. Zerze

Although some distinguishing features of RNA may be attributed to RNA's folding, the long time scale of this process is still elusive. This sluggishness can result from alternative configurations with similar free energy to the native ones. A group of tetraloops was studied in our previous work, and misfolded configurations with free energy competitive to correctly folded structures were identified, supported by experimental results. From a technical point of view, parallel-tempering in the well-tempered ensemble combined with well-tempered metadynamics (PTWTE-WTM) showed less simulation time and more reliable results compared to plain parallel-tempering (PT). In this study, we adopted an alternative protocol, namely On- the-fly Probability Enhanced Sampling (OPES) method. Comparing the two different sampling schemes showed that the multithermal-multiumbrella OPES scheme can successfully replace the PTWTE-WTM scheme and is more efficient than the PTWTE-WTM scheme.

360ax Understanding the Normal Bicontinuous Cubic Phase in Gemini Lyotropic Liquid Crystals in Order to Design Selective Separations

Nathanael Schwindt, Subin Sahu, Douglas L. Gin, Richard D. Noble, Michael Shirts

Lyotropic liquid crystal (LLC) monomers are promising materials for membrane separations due to their uniform, nanoporous structures and robust crosslinked networks. LLCs show significant variation in phase morphology, self-assembling into a number of highly ordered phases. The normal bicontinuous cubic (Q_I) phase is of particular interest for membrane separations due to its tortuous, annular pores that percolate in three dimensions. However, it is difficult to study the nanoscale structure and transport of LLC membranes experimentally.

Molecular simulation provides a route to fully explore these new membrane materials. The complexity of the Q_I phase and the long timescales required for self-assembly prevent direct atomistic simulation. However, we have developed a procedure to locate stable simulations of the Q_I phase, which includes knowledge of the ideal geometry, atomistic simulations of the lamellar bilayer, and coarse-grained models of the monomer. Studying how the atomistic model of the lamellar bilayer fits within the mathematical representations of the Q_I phase reveals the most likely structure is the gyroid *Ia3d* space group. We achieve a metastable coarse-grained model of the gyroid structure by tuning the model to match properties of the atomistic bilayer. We locate a stable atomistic gyroid structure by backmapping the coarse-grained model to an atomistic model. From the atomistic model, we characterize the local pore environment within the Q_I phase. Our simulations reveal a dynamic pore, ranging in size from 0.25 nm to 0.75 nm, which is consistent with molecular-size rejection seen experimentally. The pore consists of a percolating hydrophilic region without well-defined pore boundaries. The LLC monomers penetrate the solvent-filled region, which restricts the hydrophilic region to smaller pore diameters. Preliminary transport simulations of the crosslinked membrane show hindered water diffusion in the membrane similar to experiment.

360ay Parameterization of C, Si, and Ge in the Common Harmonic Form for Molecular Dynamics Simulations

Katarina Odak, Hendrik Heinz, Alan Weimer

The Interface Force Field (IFF) project focuses on an all-atom force field and surface model database for simulation of nanostructures of 1 to 1000 nm size, including minerals, metals, oxides, and (bio)polymers in high accuracy.[1] To achieve this aim, the IFF project develops interpretable models and parameters to simulate bulk and surface properties, including complex mixtures and reactivity (IFF-R).

In this work, we parameterized C, Si, and Ge using a common harmonic energy expression by representing the tetrahedral localized bonding with ancillary, low-mass, virtual electrons. The innovation lies in the representation of covalent-type bonds through a combination of virtual electrons and Lennard-Jones parameters. The models show some promise to describe these structures consistently. Assembling the atomic units into cubic crystals, we calculate density, melting point, constant pressure heat capacity, surface energy, and bulk modulus within 5% of experimental values. Our models accurately reproduce the bulk and surface properties of group-14 elemental crystals while retaining compatibility with existing organic and biological forcefields, allowing molecular dynamics simulations of inorganic/organic systems, mixed phases, and alloys.

References:

(1) Heinz, H., Lin, T.-J., Kishore Mishra, R., andamp; Emami, F. S. (2013). Thermodynamically Consistent Force Fields for the Assembly of Inorganic, Organic, and Biological Nanostructures: The INTERFACE Force Field. Langmuir, 29(6), 1754–1765. https://doi.org/10.1021/la3038846

360az The Role of Antiretroviral Therapeutics As Both Inhibitors and Substrates of P-Glycoprotein

Daisy Fuchs, Sahana Balaji, Kayla Sprenger

Transporting drugs across the blood-brain barrier (BBB) is a major crux in developing pharmaceuticals for treating neurological disorders and infections in the brain. Due to the difficulty of reaching effective therapeutic levels of antiretroviral (ARV) drugs in the brain, HIV is able to quickly establish an infection in the brain, leading to neurological dysfunction over time and a viral population that is difficult to ever fully eradicate. The presence of efflux proteins embedded in the endothelial cells of the BBB is one reason that crossing the BBB is such a hindrance to small molecule drug transport. The most prevalent efflux protein in the BBB is the ATPase protein, P-glycoprotein (P-gp). Protease inhibitors (PIs) that are commonly used to treat HIV are all known substrates of P-gp. However, some PIs, such as ritonavir, also behave as inhibitors of P-gp. A better understanding of the interactions between P-gp and PIs, both as substrates and inhibitors, is necessary to design ARVs that can more readily pass through the BBB and penetrate the brain microenvironment.

Molecular docking and molecular dynamics (MD) simulations are powerful and crucial computational tools for understanding how small molecule drugs, such as PIs, interact with proteins, such as P-gp. Our molecular docking results show that inhibitory PIs, such as ritonavir, bind in the ATP active site of P-gp, as well as in the substrate binding domain. The former binding location indicates that competitive inhibition is a possible mechanism of PI-based inhibition of P-gp. Comparatively, our results show that PIs that are not known to inhibit P-gp, such as saquinavir, only bind in the substrate binding domain of P-gp. MD simulations provide further insight into the free energy of binding of each PI to P-gp, and how binding of the PIs at different sites on P-gp differentially impacts the structure and dynamics of P-gp in solution. In this work, we also describe computational results can be fully understood in the context of experimental inhibition data and the bioavailability of different PIs. Overall, our molecular level insights on PI/P-gp binding may serve as a guide for designing new therapeutics that can inhibit or avoid binding to P-gp altogether, improving our ability to deliver ARVs to the brain for treating the neurological complications of HIV infection.

360ba Modeling of Photoresist Pattern Formation in EUV Lithography Through Molecular Dynamics Simulations

Seungtae Kim, Sangwoo Kwon, Won Bo Lee

As competition for micro-processes of semiconductor circuits intensifies, there is a growing need for securing a photoresist suitable for an ultra-fine line width and technology for reduction of RLS trade-off. In recent years, a polymer that introduces acid cleavable moiety (ACM) that is decomposed by reacting with acid has been actively developed. Under ultra-fine line width conditions, the radius of gyration (Rg) of each polymer chain has a size similar to that of the line width. Therefore, it is necessary to control the length of the polymer chain at the exposed/unexposed interface in consideration of chain decomposition. Thus, we explore ACM introduction strategies that can reduce the line edge roughness of the pattern while maintaining other performance properties. First of all, the Coarse-grained molecular dynamics (CGMD) simulation platform is established to precisely control various parameters such as the position of the chain scission in the mesoscopic level. This model can demonstrate the molecular weight distribution of the interface according to the presence or absence of ACM and calculate structural homogeneity of residual polymer interfaces after the process. After the principle and feasibility of the chain scission strategy are verified through CGMD simulation, all-atomic molecular dynamics (AAMD) simulation can be performed to confirm the specific interaction between the polymer resin and the developer to be modeled. An atomistic structure of the interface is back-mapped from the CGMD structure. This model verifies the interaction parameters between polymer-solvent, and the change of solubility according to the properties of ACM. Through this, we explore the optimal polymer-solvent candidate group. This approach guides further optimization to design a high-performance photoresist that overcomes the RLS trade-off by utilizing the polymer main chain scission and we anticipate that It can produce large-scale data by establishing an efficient simulation progress platform.

360bb Benchmarking Martini 3.0 Force Field for Reproducing Thermodynamic Properties of Biomolecular Condensates

Ayush Gupta, Gul H. Zerze

Multivalent and weak interactions between chains of intrinsically disordered proteins (IDPs) lead to formation of biomolecular condensates having liquid-like properties. It has been recently shown that during the transcription process, RNA Polymerase is recruited to promoter sites via a similar condensation mechanism and such transcriptional condensates have been found to have implications in cancer. Although atomistic molecular dynamics (MD) can provide high resolution insights into the liquid-liquid phase separation of IDPs, its high computational cost still requires coarse-graining the system to decrease the degrees of freedom. We aim to perform coarse-grained (CG) MD simulations of transcriptional condensates using the Martini force field. However, it has been shown that Martini 3.0, a CG force field, overestimates the protein-protein interactions. To address this issue, in this work, we rebalance the force-field using the low-sequence complexity domains of FUS and TDP-43 as a benchmarking system. By using deep learning, the obtained thermodynamic and transport properties from these simulations will be compared with experimental data to optimize the scaling parameters for the Martini 3.0 force field.

360bc Understanding Protein Unfolding Under Different Stressors from Molecular Dynamics Simulations

Yinhao Jia, Janani Sampath

Protein based therapeutics (biologics) have revolutionized the field of biopharmaceuticals and the industry is expected to grow steadily in the next decade. There are over 100 approved proteins for clinical use, including monoclonal antibodies, hormones, and fusion proteins. While proteins have a wide range of applicability and function, a major drawback is their instability, which makes them too fragile to be used alone as drugs. During storage or administration, proteins tend to unravel and lose their secondary structure due to adsorption to surfaces, interaction with other drug constituents, or changes in solution pH and temperature. Understanding how these conditions change protein structure can help with the formulation of a robust conjugate which can prevent proteins from misfolding. Quantification of weak links in the protein that are most susceptible to collapse when subjected to external stressors can help in the design of specific excipients that can prevent the degradation of these links, and aid in the rational design of conjugates that will stabilize and maintain protein structure.

To this end, we perform all-atom MD simulation to study the unfolding process of three proteins – egg-white lysozyme, insulin, and PKD Domain 1, under three different stressors – temperature, shear, and pH. Global secondary structure change during protein denaturation is characterized using backbone root-mean-square deviation (Cα-RMSD), hydrogen bond number, and hydrophobic solvent accessible surface area (SASA) and validated by using secondary structure plot and representative simulation trajectory snapshots. To further investigate specific links that breakdown first, root mean square fluctuation (RMSF), radial distribution function, and hydrogen bond formation, are systematically evaluated. Future work will focus on designing specific polymer conjugates that will prevent the onset of degradation of weak links to boost protein stability.

360bd CCR5-Eriously? Reexamining HIV-1 Tropism Switching with *In Silico* directed Evolution

Jonathan Faris, Bailey Zinger, Brenden Petersen, Kayla Sprenger

The human immunodeficiency virus type 1 (HIV-1) hijacks the CD4 receptor on a plethora of immune cells to enable viral fusion and cellular entry. HIV-1's envelope glycoprotein, Env, is comprised of the gp120 and gp41 subunits. Gp120 serves as the agonist for immune cell receptors and makes use of CD4, as well as one of two chemokine receptors, namely CCR5 or CXCR4. Which chemokine receptor gp120 utilizes defines a specific strain's tropism: R5-tropic (able to utilize CCR5), X4-tropic (able to utilize CXCR4), or R5X4-tropic (able to utilize either receptor). Transmitted/founder (T/F) strains of HIV-1—strains which can seed new infections—are overwhelmingly, if not all, R5-tropic. In late-stage HIV-1 patients, tropism switching has been observed. This phenotypic change is associated with the fast depletion of CD4+ T Cells and the rapid onset of acquired immunodeficiency syndrome (AIDS). Herein we describe the development of a computational pipeline for exploring the evolutionary dynamics that drive HIV-1 tropism switching during disease progression. Specifically, we combine an *in silico* directed evolution model adapted from a previously published model of antibody evolution in response to HIV-1 vaccination, with established machine learning algorithms and physics-based simulations. We validate our results against existing experimental datasets relating HIV-1 sequence and tropism. Ultimately, our results provide new fundamental insights into viral/immune cell receptor interactions that drive HIV-1 pathogenesis and disease progression, which may enable the design of novel strategies for harnessing the immune system to combat HIV-1 infection.

360be Monte Carlo Simulations Predicting Adsorption of 1,4-Dioxane in All-Silica Zeolites

Samiha Sharlin, Tyler R. Josephson

1,4-dioxane is used to manufacture both industrial and consumer products. Its stability, solubility, and widespread use have affected drinking water sources. Consequently, the U.S. Environmental Protection Agency has classified it as an emerging organic contaminant and a probable human carcinogen by all routes of exposure[1]. However, the incompetence of conventional remediation processes demonstrates the need for an effective treatment strategy.

Separation with nanoporous materials requires high capacity and selectivity to adsorb 1,4-dioxane preferentially and exclude water and its co-contaminants[2]. We propose to use hydrophobic all-silica zeolites with optimal pore sizes to selectively adsorb 1,4-dioxane from water. The TraPPE united-atom model for 1,4-dioxane[3] was validated by reproducing its vapor-liquid equilibrium properties using Gibbs ensemble Monte Carlo simulations, with Self-Adapting Fixed-Endpoint Configurational-Bias Monte Carlo (SAFE-CBMC) to sample the flexible ring configurations of 1,4-dioxane. This model and TraPPE-zeo[4] were used to predict 1,4-dioxane adsorption in different zeolite frameworks, investigate the adsorption mechanism, and analyze favorable host-guest interactions into zeolite channels under environmental conditions.

This material is based upon work supported by the National Science Foundation under Grant No. 2138938

[1] Adamson DT; Piña EA; Cartwright AE; Rauch SR; Hunter Anderson R; Mohr T; Connor JA. Sci Total Environ, 596-597:236-245 (2017)

[2] Chen R; Liu C; Johnson NW; Zhang L; Mahendra S; Liu Y; Dong Y; Chen M. Chemical Engineering Journal, 193-202:1385-8947 (2019)

[3] Keasler SJ; Charan SM; Wick CD; Econonmou IG; Siepmann JI, J. Phys. Chem. B 116, 11234-11246 (2012)

[4] Bai P; Tsapatsis M; Siepmann JI, J. Phys. Chem. C 117, 24375-24387 (2013)

360bf Coarse-Grained Models of Polyetherketoneketone (PEKK) Used to Perform Fusion Weld Simulations and Predict Resulting Mechanical Properties.

Chris Jones, Jenny Fothergill, Rainier Barrett, Eric Jankowski

Carbon fiber composites with a thermoplastic matrix offer manufacturing advantages over thermoset-based composites, particularly in terms of high-throughput production through fusion welding of initially separate parts. With this approach, chain diffusion and entanglement across an interface are the primary load bearing mechanisms as opposed to mechanical fasteners or adhesives. Therefore, understanding the atomistic details of polymer diffusion during fusion welding is central for understanding how processing influences structure and how that ultimately impacts mechanical strength. Molecular simulations offer insight into these details, but the long time scales relevant to polymer relaxation, entanglement, and diffusion are challenging to access. Here we implement the multi-state iterative Boltzmann inversion (MSIBI) method of coarse-graining and develop a transferable MSIBI python package to generate coarse-grained potentials for the thermoplastic matrix polymer polyetherketoneketone (PEKK). Fusion bonding simulations of PEKK are performed over a range of bulk polymer preparations and system sizes. This workflow for modeling fusion welding leverages the MoSDeF.org ecosystem of simulation tools, particularly mBuild, foyer, HOOMD-Blue and signac.

360bg Investigating Stable and Active Catalysts for Hydrogen Generation Via Methane Pyrolysis in Molten Media, Using *Ab Initio* Molecular Dynamics

Ojus Mohan, Samir H. Mushrif

Hydrogen has the potential to replace fossil fuels as a sustainable energy carrier in a variety of applications. Methane pyrolysis ($CH_{4(g)} \rightarrow C_{(s)} + 2H_{2(g)}$), enables the possibility to produce CO_2-free hydrogen from natural gas. Since methane is a highly stable molecule with strong C-H bonds, a catalyst is required to overcome kinetic limitations and achieve industrially viable reaction rates. Previous attempts to employ solid metal catalysts failed due to the catalytic deactivation caused by carbon formation on the surface and/or its diffusion into the bulk of the catalyst. Remarkably, the deactivation of the catalyst can be avoided by performing the reaction in molten salt since the carbon floats on the molten salt surface. Alternately, if C-C coupling can be facilitated, we may fully eliminate carbon formation and obtain more valuable C_2 products, along with hydrogen. However, most salts are catalytically less active compared to metals, requiring extremely high temperatures to decompose methane molecule and they cannot facilitate C-C coupling either. In this work, the objective is to enhance the catalytic performance of molten media without compromising its stability, and to facilitate C-C coupling by dispersing metal catalysts (Ni,Cu) into salt. We employed ab-initio molecular dynamics simulations combined with the accelerating approach metadynamics, which enables simulating reaction dynamics at a finite temperature and computing free energy barriers along the minimum energy pathway. We discovered that i)carbon diffused into Ni and Cu, leading to catalyst deactivation and that ii)boron doping altered the microstructure of Ni and Cu, preventing carbon diffusion into the catalyst. We further investigated C-C coupling reactions in pure and boron-doped metal-salt systems and compared the energetics for sequential methane dehydrogenation and C-C coupling. The mechanistic insights revealed in this study will aid in the development of active and stable molten media based systems for methane pyrolysis, paving the way for its commercialization.

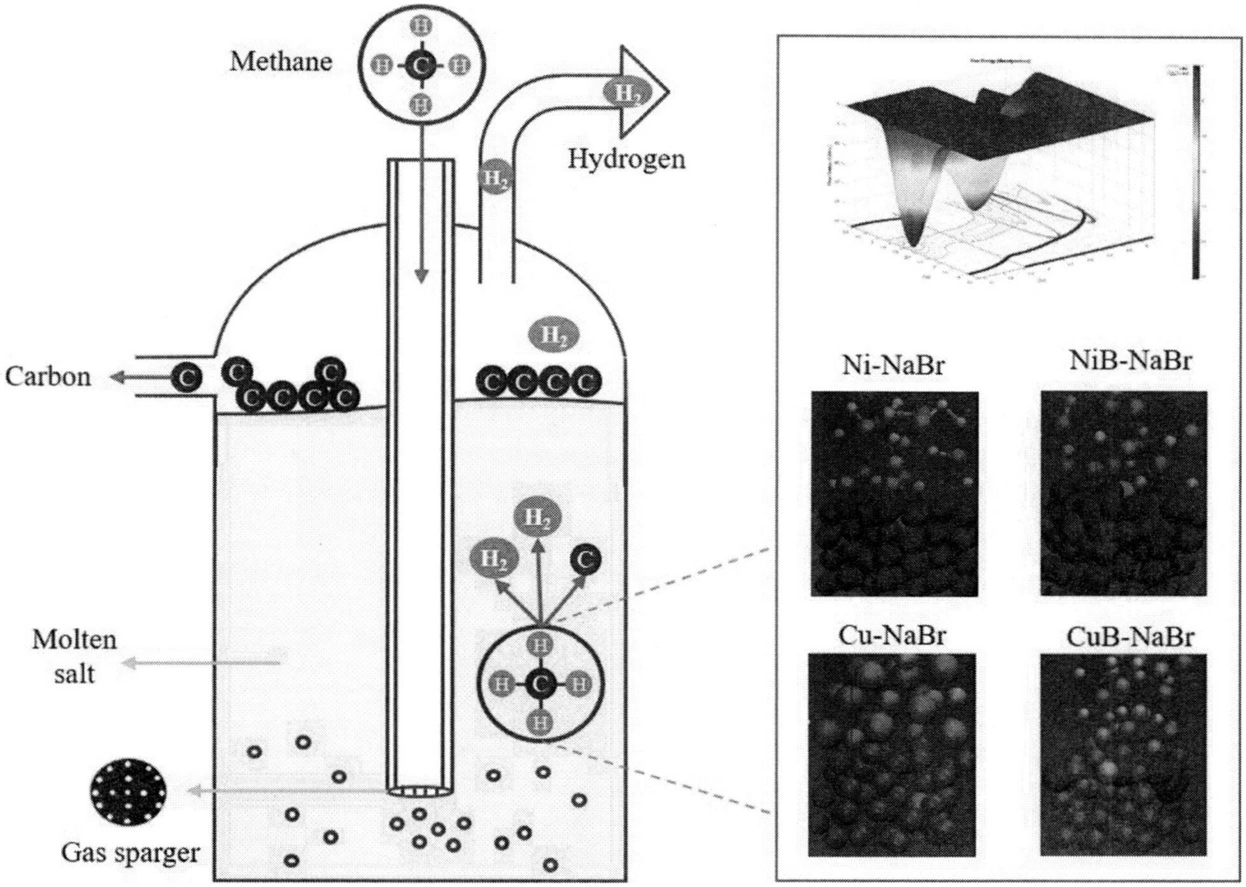

[Left] Schematic representation of methane pyrolysis in molten media. In the bubble column reactor, methane is bubbled through the melt and it decomposes to produce hydrogen and carbon. Hydrogen can be collected from the top of the reactor and carbon floats on the surface of the melt. [Right] CPMD-Metadynamics simulation system showing carbon diffusion in Ni, NiB, Cu, and CuB. Blue, orange, green, white, yellow, and pink balls represent nickel, copper, bromine, sodium, carbon, and boron respectively.

360bh Extending the MolMod Database to Transferable Force Fields

Sebastian Schmitt, Gajanan Kanagalingam, Daniel Fröscher, Florian Fleckenstein, Hans Hasse, Simon Stephan

The MolMod database [1] is an open access web-based database for classical force fields for fluids that can be used in molecular dynamics or Monte Carlo simulations. Originally, the MolMod was developed for component-specific force fields, i.e. the model parameters of a given force field are only valid for a certain component. In this work, the MolMod database was extended to transferable force fields. Therefore, a graphical user interface was implemented in the front-end of the MolMod website that enables the user constructing any organic molecule. Moreover, the user can specify, which transferable force field is to be used, e.g. TraPPE or OPLS. Then the website provides the actual force field parameters for the chosen molecule. Moreover, the MolMod database provides ready-to-use input files for different simulation engines. This enables researchers to include a large number of molecular models easily in their individual workflows.

[1] S. Stephan et al. Mol. Simul. 45 (2019) 806.

360bi Mass Transfer through Vapor-Liquid Interfaces of Binary Mixtures studied by Non-Stationary Molecular Dynamics Simulations

Dominik Schaefer, Jens Staubach, Simon Stephan, Hans Hasse

In separation processes not only thermodynamic bulk but also interfacial properties play a crucial role. In classical theory, a vapor-liquid interface is a two-dimensional object. In reality it is a region in which properties change smoothly over a few nanometers from its liquid bulk to its gas bulk value. Properties of that interfacial region can be predicted using theoretical methods, for example with molecular dynamics simulations (MD) or density gradient theory (DGT). Unexpected effects are observed in that transition region for many mixtures. While the total density changes monotonously from the bulk vapor to the bulk liquid phase, this does not hold for the molarities of the components. The molarity of the low-boiling component can have a distinct maximum at the interface, which is called enrichment [1-3]. That maximum would be an insurmountable obstacle to mass transfer according to the Fickian theory and, hence, is suspected to influence for example fluid separation processes like absorption or distillation.

To study the influence of the enrichment on the mass transfer, a new non-stationary molecular simulation method was developed in the present work, in which the penetration of a low-boiling component 2 into a liquid phase that initially only contains the pure solvent 1 is studied using a multi-ensemble technique. Five model mixtures (exhibiting different phase behavior and interfacial properties) were investigated. The results provide new insights in the physical processes involved in gas-liquid mass transfer. In particular, they indicate that the enrichment at the interface hampers the mass transfer. Furthermore, it was observed, that the penetrating
component may be repelled from the interface in early stages of the mass transfer. The results from this work complement and confirm the findings obtained earlier by our group in stationary NEMD simulations of mass transfer [4].

References

[1] S. Stephan, H. Hasse, Phys. Rev. E 101 (2020) 012802

[2] S. Stephan et al., J. Chem. Phys. 150 (2019) 174704

[3] S. Stephan, H. Hasse, Int. Rev. Phys. Chem. 39, 3 (2020) 319-349

[4] S. Stephan et al., Mol. Phys. 119, 3 (2021) e1810798

Theoretical Investigation of The Coverage Effect on Ni-In Intermetallic Catalysts for Selective Hydrogenation of Acetylene to Ethylene

Zahra Almisbaa[1], Hassan Aljama[2], Luigi Cavallo[3] and Philippe Sautet[1,4]*
[1]*Department of Chemical and Biomolecular Engineering, University of California, Los Angeles, Los Angeles, California 90095 (United States)*
[2]*Research & Development Center, Saudi Aramco, Dhahran 31311 (Saudi Arabia)*
[3]*King Abdullah University of Science and Technology, Thuwal 23955 (Saudi Arabia)*
[4]*Department of Chemistry and Biochemistry, University of California, Los Angeles, Los Angeles, California 90095 (United States)*
* sautet@ucla.edu (corresponding author)

Ethylene streams produced from naphtha steam crackers are the building blocks for polyethylene products and other key polymers. Acetylene impurities in ethylene streams impact the life span of the polymerization catalyst, and the quality of the polymeric products [1]. Catalysts are utilized to selectively hydrogenate acetylene to ethylene. The side reactions for the semi-hydrogenation of acetylene are the over hydrogenation of ethylene to ethane and the oligomerization of acetylene through C-C bond formation. Oligomeric species poison the active sites of the catalyst causing irreversible deactivation [1,2]. The deteriorating effects of these oligomers during the catalytic process intensify the demand for innovative selective solutions.

Palladium-based catalysts have been used for decades for hydrogenation reactions. The cost of Pd as well as the limited selectivity, introduce the need to find noble metal-free alternatives. Ni-based catalysts are cost-effective promising candidates, particularly the intermetallic compounds (IMC). IMC offer more control over the ensemble of active sites, improved surface dispersion, and less risk of surface segregation in comparison to disordered alloys. The Ni-In phase diagram shows that Ni_3In, Ni_2In, NiIn, and Ni_2In_3 all exist in the intermetallic phase. The promotional effect of introducing In to metallic Ni was evaluated against monometallic Ni surface. Exploring IMC with varied Ni surface content provides a better understanding of the optimum ratio of active transition metal to a second metal.

In this work, DFT-based modelling of Ni_3In, Ni_2In, NiIn, and Ni_2In_3 was used to describe the coverage effect on the activity and the selectivity of semi-hydrogenation of acetylene. When constructing traditional coverage-independent DFT-based energy diagrams, adsorption energy and activation barriers were assumed to be coverage independent or at low coverage. This means that the adsorbate-adsorbate interactions on the surface were not considered. This assumption may cause disagreement between experimental and theoretical studies and mispredict the selectivity of catalysts [2,3]. Therefore, coverage-independent and coverage-dependent reaction profiles were constructed based on the elementary steps of the hydrogenation and oligomerization reactions. Literature showed that the difference between the cross-interactions of C2 intermediates and the self-interaction was insignificant [3]. Thus, the coverage effect was approximated using a single type of species to simulate the adsorbate-adsorbate interactions on the surface and the binding energy of acetylene was evaluated at different surface coverages. Results show that the coverage effect on the energy barriers and rate-determining steps was not similar on all Ni-In surfaces. It was also shown that energy barriers were generally more processable at high coverage due to the weakened binding energies. Considering the coverage effect helps in accurately predicting the selectivity of the catalysts and provides insights to tune the reaction toward the desired product.

1. Vignola, E., Steinmann, S.N., Al Farra, A., Vandegehuchte, B.D., Curulla, D., and Sautet, P. *ACS Catal.* 8(3), 1662–1671 (2018).
2. Spanjers, C.S., Held, J.T., Jones, M.J., Stanley, D.D., Sim, R.S., Janik, M.J., and Rioux, R.M. *J. Catal.* 316, 164–173 (2014).
3. Zhao, J., Zha, S., Mu, R., Zhao, Z., and Gong, J. *J. Phys. Chem. C* 122,6005-6013 (2018).

622a Inverse Design of Pore Wall Chemistry to Control Solute Transport through Molecular Simulation Coupled Optimization

Sally Jiao, M. Scott Shell

Next-generation membranes for water treatment that move beyond simple desalination require materials with precisely tuned functionality. In particular, the purification and reuse of highly contaminated waters, such as oilfield-produced water, face key challenges in removing a wide array of solutes, including small neutral solutes that are difficult to separate. Multifunctional membrane surfaces potentially provide a vast, underexplored design space to improve membrane transport properties, but are difficult to design through trial-and-error. Here, we demonstrate an inverse design computational approach to efficiently identifying promising materials. We develop an optimization workflow coupled to molecular simulations to engineer the transport of water relative to that of boric acid in a model nanopore by spatially patterning the pore wall with nonpolar methyl and polar hydroxyl groups. The genetic algorithm optimization identifies non-intuitive functionalization strategies that hinder the transport of boric acid through the pore, simply by altering the functional group patterning. We show that a machine-learned surrogate function can predict the performance of novel patterns based only on features of the patterns and significantly accelerates the optimization. Clustering of the patterns reveals design rules which we use to create rationally designed patterns revealing the distinct mechanisms by which water and boric acid transport through the pore. Finally, we demonstrate that these inverse design techniques can also be applied to multi-objective optimization and show that the optimal patterns may help break the pernicious permeability-selectivity tradeoff in membrane design. This inverse design procedure and the accompanying molecular-level understanding of the effect of chemical heterogeneity on solute selectivity demonstrate new routes to the design of membrane materials with novel functionalities, and more broadly, show that surface chemical patterning can modulate water-mediated surface-solute interactions in systems relevant to many other technologies.

622b Multi-Multimolecular Latent Space Simulations of DNA Hairpin-Duplex Competition

Mike Jones, Andrew Ferguson

As computational resources continue to expand and force fields become more exact, molecular dynamics simulations will inevitably be applied to larger and more complex systems. Increasingly, these systems involve multiple components whose dynamics are coupled when in bound or interacting states and are otherwise entirely independent. Traditional kinetic modeling methods such as Markov state models (MSMs) have been applied to these multi-molecular systems but can fail to capture an exponentially increasing number of often degenerate states. We previously developed latent space simulators (LSS) as a continuous approach to low-dimensional kinetic modeling that (i) encode a molecular system into a slow latent space, (ii) propagate dynamics in this latent space, and (iii) generatively decode a synthetic molecular trajectory. Here we expand on the LSS framework to develop a model, specialized toward multi-multimolecular systems, which learns the dynamics of each molecule independently and varies the reconstruction approach as a function of intermolecular distance and orientation. We apply our model to a coarse-grained DNA system consisting of oligomers that can both hybridize into a duplex and fold into unimolecular hairpins. Our model generates ultra-long synthetic molecular trajectories that reproduce the thermodynamics and kinetics of molecular dynamics data and lends insights into competition between duplex and hairpin states. This work lays the foundations of the approach for applications to other multi-molecular systems such as protein dimerization, DNA-protein binding, and protein-ligand interactions.

622c Machine Learning-Enabled Prediction of Electronic Properties of Radical Polymers at Coarse-Grained Resolutions

Riccardo Alessandri, Juan J. de Pablo

Non-conjugated radical polymers are conducting polymers with non-conjugated backbones bearing pendant stable radical sites. These polymers promise sustainable and tunable materials for applications in, for example, all-organic energy storage devices. However, a fundamental understanding of how the chemical structure impacts charge transport in these materials is still lacking. To address this, efficient computational schemes that take into account both electronic properties and mesoscale morphological features are required.

We present an efficient computational scheme that uses supervised machine learning (ML) to predict electronic-structure information relating to charge transport at coarse-grained (CG) resolutions. ML models are trained on data obtained from quantum chemical calculations on all-atom conformations sampled from condensed-phase simulations. Predictions of several electronic properties are investigated as a function of CG resolution. Trained ML models then enable electronic property predictions directly from CG polymer morphologies. We evaluate both systematic and building-block coarse-graining techniques. We validate the approach by comparing to state-of-the-art methodologies that require backmapping to atomistic resolution and subsequent quantum chemical calculations. The ML-assisted scheme shows potential for drastically accelerating multiscale computational workflows that connect electronic properties to the mesoscale, thereby enabling high-throughput modeling aimed at the understanding and design of radical polymers and other soft electronic materials.

622d Machine Learning Guided Discovery of Polymer Membranes for Reducing Greenhouse Gas Emissions

Yasemin Basdogan, Zhen-Gang Wang

Developing polymer membranes which have high gas permeability and selectivity remains a grand challenge for energy, the environment, and economic sustainability. Our main objective with this project is designing polymer membranes with optimal permeability and selectivity by Machine Learning (ML) assisted polymer engineering and high throughput screening. We hypothesized that ML models can accurately predict gas permeability and selectivity of both existing and hypothetical membranes when trained on large and diverse data. First, we created a library of polymers and studied them in detail by calculating their physicochemical properties. Next, we employ unsupervised ML to study the structural similarities of polymers and map their known materials space. Then, we conduct a feature selection protocol to pick the optimal features that correctly represent the similarities and differences of our polymer materials. Once the features are identified, we use them to train an ensemble of different ML models. Using appropriate validation techniques, we select the top-performing ML model that can correctly predict gas diffusivity in polymer membranes. Later, we will use a Genetic Algorithm (GA) – guided by our trained ML model – to effectively sample the polymer materials space and identify high performance polymers for CO_2/N_2 separation. Finally, we will use experiments to validate the new materials identified by this procedure. This work delivers a robust hypothetical polymers database and a computational framework to further explore the polymer materials space for any gas separation problem of interest.

622e Sigma Profiles in Deep Learning: Towards a Universal Molecular Descriptor

Dinis O. Abranches, Yong Zhang, Edward Maginn, Yamil Colón

The ability of deep neural networks (DNNs) to correlate variables whose relationship is unknown or too complex to be derived is attracting a great deal of interest in chemistry-related fields. However, most DNN-compatible molecular representations commonly used, namely string-based vectors (SMILES or SELFIES), molecular fingerprints, molecular graphs, and Coulomb matrixes, present several shortcomings. In particular, they encapsulate little chemical information beyond atom type and connectivity and their size depends on the size of the molecule: the larger the molecule, the larger the vector or matrix used to represent it. Thus, the input size of a DNN must be made as large as the largest molecule available in the dataset of interest. This leads to the development of complex DNNs that possess many trainable parameters and, thus, need very large datasets to be properly fitted.

By virtue of being unnormalized histograms of screened charges, σ-profiles encode a great deal of chemical information (charge density, polarity, etc.) and their size does not change with the size of the molecule, mitigating the disadvantages of the descriptors mentioned above. As such, this work showcases, for the first time, the ability of σ-profiles to function as universal molecular descriptors in deep learning. To do so, the σ-profiles of 1432 compounds were used to train convolutional neural networks (CNNs) that accurately correlate and predict a wide range of physicochemical properties (molar masses, normal boiling temperatures, vapor pressures, densities, refractive indexes, and aqueous solubilities). To boost their performance, the architecture and hyperparameters of each CNN were optimized using a battery of algorithms, particularly Bayesian Optimization and Local Search. Furthermore, it was shown that thermodynamic conditions, namely temperature, can also be used as additional inputs to broaden the applicability of the models. Among all other advantages mentioned, this work shows that σ-profiles can extend the use of deep learning methodologies to areas where datasets are relatively small and scarce.

622f Systematic Investigation of Training Protocols for Machine Learning Derived Interatomic Potentials

Nisarg Joshi, Jim Pfaendtner

Molecular simulations have proved to be an important tool in understanding molecules and molecular interactions and have helped accelerate research in material discovery, catalysis design and biomedicine. Molecular dynamics (MD) simulations help understanding the conformational changes of molecules from the atomic configurations and obtaining potential energy surface (PES) for the systems. However, there is always a tradeoff between the accuracy and computational cost. While *ab initio* molecular dynamics (AIMD) offer improvements in the accuracy of results and preclude the need to pre-define molecular topology, poor scalability of electronic structure methods severely limits applicable system sizes. On the other hand, classical molecular dynamics methods significantly improve scalability, but are limited to low accuracy and other well-known limitations. To bridge this gap, there have been developments in using machine learning (ML) methods to develop an interatomic potential for molecules[1]. Typical strategies seek to create ML potentials trained on AIMD data, which offer greater accuracy[2] while also being able to simulate larger, more relevant systems. However, there is a bottleneck in developing ML potentials that surrounds the breadth of required training data[3] needed to create sufficient reconstruction accuracy of energies and forces. To overcome this challenge, we have explored using enhanced sampling methods to speed the generation of training data and improve the reconstruction accuracy of the potentials while also limiting the total amount of training data needed. We envision the applications of ML potentials on ion interactions with solvents involving long range interactions and on different ranges of concentration. We also illustrate the generalizability of ML potentials for liquids and explore the extent to which they can be used for systems of varied size and different thermodynamic ranges.

References:

1. Behler J, Parrinello M. Generalized Neural-Network Representation of High-Dimensional Potential-Energy Surfaces. *Phys Rev Lett.* 2007;98(14):146401. doi:10.1103/PhysRevLett.98.146401
2. Unke OT, Chmiela S, Sauceda HE, et al. Machine Learning Force Fields. *Chem Rev.* Published online March 11, 2021. doi:10.1021/acs.chemrev.0c01111
3. Vassilev-Galindo V, Fonseca G, Poltavsky I, Tkatchenko A. Challenges for machine learning force fields in reproducing potential energy surfaces of flexible molecules. *The Journal of Chemical Physics.* 2021;154(9):094119. doi:10.1063/5.0038516

622g Application of Machine Learning to Accelerate High-Throughput Molecular Dynamics Screening: A Study of Tribological Properties of Monolayer Films

Co D. Quach, Justin Gilmer, Daniel Pert, Akanke Mason-Hogans, Peter Cummings, Clare McCabe

Monolayer films have shown promise to reduce friction and wear of mechanical devices with separations on the nanoscale.[1] These films have a vast design space owing to the many tunable properties that have been shown to impact tribological behavior, including terminal group chemistry, film thickness and film composition. Here, the Molecular Simulation Design Framework (MoSDeF)[2] is used to perform a combinatorial screening study of 9747 unique multi-component monolayer films (116,964 total simulations) to identify which terminal groups combinations and compositions lead to favorable tribological properties, i.e., those that minimize the coefficient of friction and adhesion forces. A random forest regressor machine learning framework[3] is developed and used to establish a more robust understanding of the connections between terminal group chemistry and tribological properties. The impact on the predictive capacity of the size/scope of the training set used to develop the machine learning models is also investigated. Here we present heuristics for determining the accuracy and efficiency of ML models for use in prescreening. The machine learning model was subsequently applied to screen 193,131 unique film candidates identified from the ChEMBL small molecule library, resulting in approximately a 5 order of magnitude speed-up in analysis compared to simulation alone. This work demonstrates the promise of computational screening, in combination with machine learning, to greatly increase the throughput in combinatorial approaches.[4] The simulations reported follow guidelines suggested by the TRUE standard, emphasizing the reproducibility and extensibility of the study.[5]

References

[1] N. S. Tambe and B. Bhushan, "Nanotribological characterization of self-assembled monolayers deposited on silicon and aluminium substrates", Nanotechnology, **2005**, 16, 1549–1558.

[2] "MoSDeF" [Online]. Available: https://mosdef.org

[3] A. Z. Summers, J. B. Gilmer, C. R. Iacovella, P. T. Cummings and C. McCabe, "MoSDeF, a Python Framework Enabling Large-Scale Computational Screening of Soft Matter: Application to Chemistry-Property Relationships in Lubricating Monolayer Films", J. Chem. Theory Comput., **2020**, 16, 1779–1793.

[4] C. D. Quach, J. B. Gilmer, D. Pert, A. Mason-Hogans, C. R. Iacovella, P. T. Cummings and C. McCabe, 'High-Throughput Screening of Tribological Properties of Monolayer Films using Molecular Dynamics and Machine Learning', J. Chem. Phys. **2022,** in press 10.1063/5.0080838.

[5] M. W. Thompson, J. B. Gilmer, R. A. Matsumoto, C. D. Quach, P. Shamaprasad, A. H. Yang, C. R. Iacovella, C. McCabe and P. T. Cummings, "Towards molecular simulations that are transparent, reproducible, usable by others, and extensible (TRUE)", Mol. Phys., **2020**, 0, e1742938.

622h Detecting Transition Boundaries in Molecular Simulations

Brandon Butler, Domagoj Fijan, Sharon C. Glotzer

In computational studies of self-assembly, protein folding, and other molecular phenomenon, a key point of interest is the transition from one state to another. The detection of such transitions within a simulation are based on order parameters tuned to the particular phenomenon, but when the possible transition states are not known, an order parameter approach is difficult to use. Furthermore, defining the beginning of a smooth transition through a threshold as is often done is sub-optimal as either it fails to capture the beginning of an event or is too sensitive to random fluctuations within the system. We present in this talk an algorithmic method for determine the number, extent, and details of a transition using change point detection. Likewise, we will showcase multiple example phase transitions highlighting the performance and limitations of our method. The systems we explore will vary from hard particle to soft colloidal to atomic systems. We also will expound on potential applications of this technique in applications such as machine learning and active learning. By partitioning the system, machine learning can preferentially learn on transition states avoiding potential over-fitting on the long periods of stability common in simulations. We will also show how this can be used as a more traditional tool for computational scientist in exploring new systems.

Adiabatic motion and statistical mechanics via mass-zero constrained dynamics

Giovanni Ciccotti

In recent work [Coretti et al., J. Chem. Phys., 2018, 149, 191102], a new algorithm to solve numerically the dynamics of the shell model for polarization was presented. The approach, broadly applicable to systems involving adiabatically separated dynamical variables, employs constrained molecular dynamics to strictly enforce the condition that the force on the fast degrees of freedom, modeled as having zero mass, is null at each time step. The algorithm is symplectic and fully time reversible, and results in stable and efficient propagation. In this paper we complete the discussion of the mechanics of mass-zero constrained dynamics by showing how to adapt it to problems where the fast degrees of freedom must satisfy additional conditions. This extension includes, in particular, the important case of first principles molecular dynamics. We then consider the statistical mechanics of the mass-zero constrained dynamical system demonstrating that the marginal probability sampled by the dynamics in the physical phase space recovers the form of the Born–Oppenheimer probability density. The effectiveness of the approach and the favorable scaling of the algorithm with system size are illustrated in test calculations of solid Na via orbital-free density functional dynamics.

383a Computational Design of Peptides As Biomarkers, Sensors, Detectors and Drugs

Carol Hall

We describe our efforts to develop an efficient computational algorithm that searches for peptides that bind strongly and selectively to specific biomolecular targets, and to use that algorithm in the design of peptide-based sensors, drugs and affinity chromatography ligands. The algorithm is an iterative procedure that involves as many as 50,000 sequence mutation moves and/or peptide backbone conformation moves to arrive at the peptide sequence and conformation that has the lowest binding energy to the target. The top scoring peptides are then further evaluated by performing explicit-solvent atomistic simulations of the peptide–target complex to determine their binding free energies. The biomarker chosen for initial study was Cardiac Troponin I, a 210-amino acid protein that is elevated in the blood of heart attack victims. The top-scoring peptides were synthesized by collaborators at the Air Force Research Laboratory who then used a variety of experimental techniques to assess the binding capabilities of the top candidate peptides. The measured binding affinity of the designed variant, P2, for cTnI was extremely strong (K_D = 0.27 nM) and comparable to that of the natural antibody (K_D = 0.12 nM). We further describe two other projects: (1) design of peptide affinity ligands for bioseparating monoclonal antibody, IgG, from the cell cultures in which they are synthesized, and (2) design of peptides to block the action of the toxins secreted by C-difficile bacteria in the large intestine.

383c Capillary Condensation and Hysteresis in Nanoporous Materials: New Simulations and New Insights

Zhao Li, Randall Snurr

Capillary condensation and hysteresis are important phenomena in the field of adsorption. Pioneering work by Keith Gubbins and his colleagues demonstrated the power of grand canonical Monte Carlo (GCMC) simulations and classical density functional theory to study hysteresis and capillary condensation and provided important insights into these phenomena. Here, we build upon this work by applying molecular simulation to study capillary condensation and hysteresis of C1 to C6 n-alkanes in the complex pore structures of metal-organic frameworks (MOFs). Two complementary simulation approaches were employed. First, we used GCMC simulations and obtained the hysteresis loops (if any) by performing the simulations sequentially "up and down" the isotherm, where the final configuration from a given fugacity is used as the initial configuration for the simulation at the next fugacity. In the second approach, we used canonical ensemble simulations at fixed loadings combined with Widom insertions to determine the fugacities. Note the complementary nature of the approaches: GCMC sets the fugacity and calculates the loading, while canonical Widom insertions set the loading and calculate the fugacity. We show that the canonical Widom approach allows us to access the full van der Waals (vdW) loop for the isotherm. GCMC simulations can generate hysteresis loops, but they may suffer from sampling inefficiencies: at high loading, GCMC insertion and deletion moves have low acceptance rates due to the high density in the pores. A canonical simulation avoids these inefficiencies by fixing the number of molecules in the simulation box. The Widom method allows us to access meta- and unstable thermodynamic states (loadings) and observe bubble formations that are impossible to capture using GCMC. Canonical isotherms also allow us to interpret and explain hysteresis loops and step changes given by GCMC simulations. Using these methods, we were able to characterize and analyze complicated phase behaviors of alkanes in complex MOF structures.

383d Robust Phase Equilibria in Reactive Non-Ionic and Ionic Systems Relevant to Pharmaceutical Applications

Amparo Galindo

Based on a concept commonly applied in mathematical programming known as duality, Mitsos and Barton [1] presented a consistent formulation for Gibbs free energy minimization as a dual extremum principle (DEP) in thermodynamics to identify points that satisfy the necessary and sufficient conditions for thermodynamic stability. Following their work, Pereira et al. [2],[3] extended the DEP framework to the volume-composition space, using the Helmholtz free energy and the relationship between optimality conditions and pressure. This approach is known as the HELD (Helmholtz free energy Lagrangian dual) algorithm and allows for the phase equilibrium problem to be solved rigorously without assumptions of the number of phases or initial guesses of the phase compositions. However, the original formulations do not account for ionic species or chemical reactions in any of the phases at equilibria. I will discuss the introduction of additional constraints in the minimization of the Gibbs free energy, such as macroscopic electroneutrality and suitable relationships between molar amounts and extents of reaction using transformed coordinates [4], that restrict the solution space to compositions that are at chemical equilibria. The extended algorithm and is tested in combination with the SAFT-γ Mie group-contribution equation of state [5] to model complex non-ionic and ionic reactive systems with industrial relevance. For instance, dissolved salts can significantly affect the phase equilibria of mixed-solvent solutions. Depending on the nature of ions and organic compounds present phenomena such as salting-out and salting-in lead to dramatic changes in solubility. In the same context, one of the more intriguing examples of water's effects on biological molecules is the zwitterion formation of naturally occurring amino acids. α-amino acids exist as zwitterions in aqueous solutions, with the neutral form not present to any measurable degree; they contain both positive and negative charged groups but with an overall neutral charge (no net charge). At high or low pH values they behave as weak acids or weak bases, respectively, leading to the formation of charged species with a remarkable effects on the activity of the solutions. I will present results of the solubility of organic solvents (e.g., alcohols, ketones, and organic acids) in water and to the corresponding changes in their phase behaviour on the addition of mono and divalent alkali halide salts, and will describe the thermodynamic behaviour of α-amino acids (e.g., glycine, alanine) as zwitterions, predicting the distribution of species as a function of pH.

[1] Mitsos, A. and Barton, P.I., 2007. A dual extremum principle in thermodynamics. AIChE Journal, 53(8), pp.2131-2147.
[2] Pereira, F.E., Jackson, G., Galindo, A. and Adjiman, C.S., 2010. A duality-based optimisation approach for the reliable solution of (P, T) phase equilibrium in volume-composition space. Fluid Phase Equilibria, 299(1), pp.1-23
[3] Pereira, F.E., Jackson, G., Galindo, A. and Adjiman, C.S., 2012. The HELD algorithm for multicomponent, multiphase equilibrium calculations with generic equations of state. Computers andamp; chemical engineering, 36, pp.99-118
[4] Ung, S., andamp;amp; Doherty, M. F. (1995). Theory of phase equilibria in multireaction systems. Chemical Engineering Science, 50(20), 3201-3216.
[5] Eriksen, D.K., Lazarou, G., Galindo, A., Jackson, G., Adjiman, C.S. and Haslam, A.J., 2016. Development of intermolecular potential models for electrolyte solutions using an electrolyte SAFT-VR Mie equation of state. Molecular Physics, 114(18), pp.2724-2749.

383e Charting the Design Space of Chemically-Heterogeneous Surfaces That Manipulate Water-Mediated Interactions

M. Scott Shell

Water-mediated interactions constitute fundamental driving forces in a profound range of synthetic and natural materials. While such interactions have now been studied extensively at a theoretical level for well-defined surface chemistries (e.g., ideal hydrophobes, simple ions, model colloids, self-assembled monolayers, etc.), the picture for highly heterogeneous solutes and surfaces that present multiple chemistries and interaction types remains far from well-understood. In this talk, I discuss the application of molecular simulations that are coupled to optimization andamp; machine-learning algorithms to systematically characterize and design within the enormous space of chemically heterogeneous surfaces. Our work specifically addresses surfaces that systematically manipulate both solute affinity and transport properties to suggest design principles for next-generation water purification membranes.

383f Wetting and Electrowetting of Polyelectrolyte Complex Coacervates on Solid Surfaces

Christopher Balzer, Pengfei Zhang, Zhen-Gang Wang

The wetting behavior of complex coacervates underpins their use in many emerging applications of surface science, particularly wet adhesives and coatings. Many factors dictate if a coacervate phase will condense on a solid surface, including solution conditions (i.e. salt or pH), the nature of the polymer-substrate interaction, and the underlying supernatant-coacervate bulk phase behavior. In this work, we use a simple inhomogeneous mean-field theory to study the wetting behavior of complex coacervates on solid surfaces both off-coexistence (wetting transitions) and on-coexistence (contact angles). We focus on the effect of salt concentration, the polycation/polyanion surface affinity, and the applied electrostatic potential on the wettability. We find that the coacervate generally wets the surface via a first order wetting transition, which evolves to a second order transition upon approaching the bulk critical point. Applying an electrostatic potential to a solid surface always improves the surface wettability when the polycation/polyanion-substrate interaction is symmetric. For asymmetric surface affinity, the wettability has a nonmonotonic dependence on the applied potential that depends on the degree of asymmetry. We use simple scaling and thermodynamic arguments to explain our results.

383g Free Energy Barriers for Anti-Freeze Protein Engulfment: A Model for the Effects of Supercooling, Footprint Size, and Neighbor Distances

Hossam Farag, Baron Peters

Biological anti-freeze proteins (AFPs) protect organisms at freezing conditions by attaching to the ice surface and arresting its growth. In the theory of DeVries et al., each adsorbed AFP locally pins the ice surface, resulting in a metastable dimple for which the interfacial forces counteract the driving force for growth. As supercooling increases, these metastable dimples become deeper until metastability is lost in an engulfment event where the ice irreversibly swallows the AFP. Engulfment resembles nucleation in some respects, and this paper develops a model for the critical profile and free energy barrier for the engulfment process. Specifically, we develop a variational model for the ice-water interface and obtain estimates of the free energy barrier as a function of the supercooling, the AFP footprint size, and the distance to neighboring AFPs on the ice surface. We discuss how this information might be used to predict thermal hysteresis properties as a function of AFP properties and AFP coverage.

541a Understanding and Controlling Properties of Biomimetic Polymers

Jim Pfaendtner

Peptoids, or n-substitued glycines, are complex and diverse oligomeric structures which have been explored for a number of biomimetic applications including drug delivery, surfactants, and catalysts. In contrast to their peptide counterparts, on peptoids the sidechain is bonded to the backbone nitrogen resulting in a flexible omega backbone dihedral that is able to isomerize into both stable *cis*- and *trans*- backbone conformations. This unique feature of peptoids allows for these structures to potentially span a significantly larger configurational space of chemical and structural functionality through the careful tuning of their side chains. This vast chemical and structural space has created significant challenges for rational design of new structures and functions as the underlying molecular scale driving forces that give rise to sequence/structure/function relationships have proven difficult to uncover. This talk will highlight recent developments from our group in the use of statistical mechanical tools to accelerate molecular simulations of rare events like peptoid folding, aggregation and adsorption on inorganic surfaces.

541b The Influence of Halides on the Solution-Phase Growth of Cu Nanowires and Microplates: A Multi-Scale Theoretical Study

Kristen Fichthorn

In experimental studies, chloride promotes the growth of Cu nanowires, with {100} sides and {111} ends, in the presence of hexadecylamine (HDA) capping agent, but iodide promotes the growth of microplates with {100} sides and large, basal {111} facets. We used *ab initio* thermodynamics based on quantum density functional theory (DFT) to understand the role of chloride and iodide in these syntheses. Our studies reveal that HDA forms self-assembled monolayers (SAMs) on both surfaces at sufficiently low halide solution-phase chemical potentials. As the chloride chemical potential is increased, we find a range where HDA is displaced from {111} facets but remains strongly bound to {100} facets. These features suggest Cu ions are easily deposited on the ends of growing nanowires, but that deposition is suppressed on the nanowire sides and are consistent with the experimentally observed growth of Cu nanowires. Additionally, adsorbed chlorine reverses the facet preference of Cu atom binding and surface diffusion from the bare Cu surface, such that Cu binding is stronger, and diffusion is slower on the {111} nanowire ends than on the {100} sides. Calculations using the theory of absorbing Markov chains indicate that a combination of facet-selective surface diffusion and deposition leads to the growth of nanowires with aspect ratios over 1000.

Unlike chloride, iodide displaces HDA from both Cu(100) and Cu(111) when it is present at relatively low solution-phase concentrations. Additionally, iodide enhances the binding and surface-diffusion tendencies of bare Cu(100) and Cu(111), such that Cu atoms are basically immobile on the Cu(100) nanoplate sides and they are highly mobile on the basal {111} facets. Calculations using the theory of absorbing Markov chains indicate that facet-selective surface diffusion leads to the growth of microplates with dimensions in the experimental range.

Overall, the profound effect of halide adsorption in promoting/suppressing molecular binding on metal surfaces extends beyond Cu surfaces and I will discuss examples from our very recent work, in which similar scenarios occur on Pd, Au, and Au surfaces.

1. J. Kim, S. Alvarez, Z. Chen, K. A. Fichthorn, and B. J. Wiley. *J. Am. Chem. Soc.* **140**, 14740-14746 (2018).
2. A. Fichthorn and Z. Chen. *J. Vac. Sci. Technol. A* **38**, 023210 (2020).
3. J. Kim, M. A. Cruz, Z. Chen, M. Brown, K. A. Fichthorn, and B. J. Wiley. *Chem. Mater.* **33**, 881–891 (2021).
4. Kim, J. Cui, and K. A. Fichthorn, *ACS Nano* **15**, 18279-18288 (2021).
5. Kim and K. A. Fichthorn, *Faraday Discussions*, DOI: https://doi.org/10.1039/D1FD00091H

541c Mechanochemical Activation in Nanostructured Triblock Copolymers: A Computational Study

Antonia Statt

Force-driven chemical reactions have emerged as an attractive platform for diverse applications in polymeric materials. However, the microscopic chain conformations and topologies necessary for efficiently transducing macroscopic forces to the molecular scale are not well-understood. In this talk, we demonstrate the importance of polymer conformations and nano-scale morphology for force-responsive copolymers by investigating the impact of network-like topology on mechanochemical activation in self-assembled triblock copolymers. Using a simple coarse-grained model, we find that activation during tensile deformation depends strongly on both the polymer composition and chain conformation in these materials, with activation requiring higher stress in materials with a higher glassy block content, and most activation occurring in the tie chains connecting different glassy domains. Additionally, we observe a spatial pattern of activation which appears to be tied to distortion of the self-assembled morphology. Higher activation is observed in the tips of the chevrons formed during deformation of lamellar samples, as well as in the centers between the cylinders in the cylindrical morphology. Our work shows that changes in the network-like topology of chains in different morphologies significantly impact mechanochemical activation efficiencies in these materials, suggesting that this area will be a fruitful avenue for further research.

541d Modeling the Recycling of Polymer Waste

Sanat K. Kumar

The fundamental issue underlining the difficulty of recycling polymers is that the free energy of polymerization is strongly favorable – thus, reversing this process to create monomers is energetically unfavorable, and hence economically not viable. A metric that characterizes this difficulty is the ceiling temperature, $T_{ceiling}$ – above $T_{ceiling}$ depolymerization processes are thermodynamically favorable. Thus, polyethylene with $T_{ceiling} \sim 610^0C$ is hard to recycle while polyethyleneterephthalate with $T_{ceiling} \sim 300^0C$ is much easier. We shall thus discuss different strategies that can facilitate this key process for families of polymers – in all cases our approach is to combine numerical simulations with back of the envelope calculations to understand experimental results. The ultimate goal is to devise family specific schemes that can bring us closer to the ideal of a circular carbon economy.

541e How Molecular Simulations Are Being Used to Phase out High Global Warming Potential Hydrofluorocarbon Refrigerants

Edward Maginn

Co-authors: Ning Wang, Ryan Smith, Eliseo Marin-Rimoldi, Bridgette Befort, Alexander Dowling, and Edward J. Maginn
Department of Chemical and Biomolecular Engineering, University of Notre Dame, Notre Dame, IN 46556

The Montreal Protocol of 1987 phased out chlorofluorocarbon (CFC) refrigerants because of their high ozone depletion potential. The replacements, typically mixtures of hydrofluorocarbons (HFCs), are safe for the Earth's ozone layer, but most have been found to have high global warming potentials (GWPs). As a result, 197 countries signed the Kigali agreement in 2016 to phase out the use of high-GWP HFCs with the goal of reducing emissions by 80% in the next 20 years. The U.S. Environmental Protection Agency is proposing its first rule under the American Innovation and Manufacturing (AIM) Act of 2020 to phase down the production and consumption of hydrofluorocarbons (HFCs). There are currently millions of metric tons of high-GWP refrigerants that will need to be reclaimed, but there are no good methods for separating and recycling the individual components, given that many are azeotropic or near-azeotropic mixtures.

In this talk, we will describe our recent work using molecular simulations to compute key thermophysical properties of HFCs and their mixtures with ionic liquid solvents and adsorbents. The simulations provide important data that can help identify technologies that can be used to recover and repurpose HFCs. Using machine learning and optimization approaches, we show how accurate HFC force fields can be developed and used to make property predictions. We describe ways of rapidly estimating low pressure Henry's Law constants of HFCs in ionic liquid solvents, enabling us to screen over hundreds of potential ionic liquid solvents. We also demonstrate how advanced free energy sampling methods can be used to compute full isotherms of HFCs and HFC mixtures in solvents and porous solid adsorbents. Finally, we use a range of equilibrium and non-equilibrium molecular dynamics simulations to compute key transport properties of HFCs and their mixtures with ionic liquids.

431a A Quantitative, Molecular Simulations Based, Kinetic Model of Acid Gas Absorption in Aqueous Amines

Frederick de Meyer, Bénédicte Poulain, Xavier Rozanska

The removal of acid gases like CO_2 and H_2S is an essential process in a natural gas plant. The state-of-the-art technology for acid gas treatment is the chemical absorption with aqueous alkanolamines. The same technology is also increasingly used to capture CO_2 from flue gases in the fight against climate change. Despite this technology existing for almost a century, many basic questions remain unanswered, preventing major progress in identifying more efficient solvents for acid gas removal and CO_2 capture.

It would, for example, be efficacious to have a quantitative and predictive model of the rate-controlling processes of acid gas absorption by aqueous tertiary amines. Despite numerous attempts to date, there is no quantitative relationship between the amine structure and the rate of absorption of acid gases (CO_2 and H_2S).

The present computational study achieves this goal by focusing on the reaction of CO_2 and H_2S with OH^- forming HCO_3^- and HS^-. The performance of the resulting model is demonstrated for a consistent experimental dataset of the absorption rates of CO_2 and H_2S for a dozen of amines. The key to the new model's success is that the reaction free energy barriers are evaluated from the solvation free energies of the reactants and the products, obtained from molecular dynamics simulations. Those simulations include subtle solvation effects of reactants and products, which turn out to be essential to understand the kinetic properties of the amines.

[1] Rozanska, X., Wimmer, E. andamp; de Meyer, F. Quantitative Kinetic Model of CO_2 Absorption in Aqueous Tertiary Amine Solvents. *J. Chem. Inf. Model.* 61, 1814–1824 (2021)

431b Accessible DPD Models for Structured Formulations Via Automated and Atomistically-Driven Parameterization

David A. Nicholson, John C. Shelley, Andrea Browning, Mayank Misra, Mohammad Atif Faiz Afzal, Paulo G. M. Mileo, H. Shaun Kwak, Mathew D. Halls

In many formulations, structure is a focal point for design and optimization. Depending on the application, structure can take on different forms–to name a few: emulsified droplets, microphases, micelles and amorphous dispersions. The fine structural detail in a formulated material can be impractical to characterize in the lab, yet essential to its functionality. With suitable models, a computer-aided design approach can facilitate design for these important details that are not otherwise accessible. The formation of structure, by and large, occurs on a spatiotemporal range that is impractical for computational modeling using all-atom simulation. As such, coarse-grained (CG) simulation, in particular dissipative particle dynamics (DPD), is an attractive approach to access the operative time- and length-scales for structured formulations.

The primary barrier to wider utilization of DPD and other CG methods is the effort required to develop a new model from scratch. In this work, we employed a recently-developed tool to make DPD modeling more accessible by automating model parameterization using all-atom simulation. We have used this tool to build models for various types of structured formulations including surfactant micelles, nano-emulsions and microphase-separated copolymers. Using these models, we will show how changes to composition, molecular architecture and chemistry affect the resulting structures. The ability to systematically develop models that are able to capture these relationships in structured formulations represents a significant advancement in the computational design of these materials.

431c Multi-Scale Simulations Relevant for Hydrates Management

Alberto Striolo, Alberto Striolo

Many academic computational studies are conducted on homogeneous, perfect systems; when heterogeneous systems are studied, atomically flat idealized surfaces are often modeled. However, most practical systems show surfaces that are chemically heterogeneous and atomically rough. How can we extrapolate the simulation results to provide useful insights for applications? In this presentation some examples from our recent research will be presented, together with their possible applications in the materials and energy sector.

Focus will be on the formation of clathrate hydrates in oil andamp; gas pipelines, which could block and sometimes rupture the pipelines. We studied molecularly thin films of surfactants adsorbed on methane hydrates. In some cases, we observed a 'frozen interface'. Interestingly, the surfactants that showed such structure seem to be effective in preventing the formation of hydrate plugs, as shown by experiments. We then conducted a variety of simulations to uncover interfacial properties of hydrate systems that could be relevant for flow assurance. As an example, we were able to correlate the molecular properties of surfactants to their ability to prevent the formation of hydrate plugs, and we achieved excellent agreement with experiments conducted with the micro mechanical force apparatus by our collaborators at the Colorado School of Mines (i.e., the group led by Prof. Carolyn Koh).

Are these results just nice pictures, or could they help trigger innovations, e.g., in materials discovery? To build a bridge towards application, we are using molecular simulation results to parameterize a stochastic model for hydrate growth, which is being developed in collaboration with Dr Michail Stamatakis of University College London.

431d Progress Towards Alchemical Simulations Driven By Many-Body Gradient-Domain Machine Learning

Alex M. Maldonado, John Keith

Alchemical predictions of molecular properties, such as solvation free energies and pK_as, provide an immense value for understanding chemical and physical processes. However, simulation convergence requires substantial sampling along the alchemical pathway, which is typically only possible with classical potentials. Nowadays, machine learning (ML) potentials can drive molecular simulations with quantum chemical accuracy at a significantly reduce cost—albeit a couple times slower than classical potentials. Training ML potentials often requires a sizable amount (e.g., tens of thousands) of quantum chemistry calculations that diminishes the overall computational cost savings. Many-body gradient-domain machine learning (mbGDML) is a recently developed framework that incorporates the size transferability of many-body expansions with the accuracy and efficiency of GDML potentials. After training, we have n-body potentials that accurately reproduce quantum chemical forces and energies with only 1000 training data points. Here, we describe the mbGDML approach and its use in predicting solvation free energies with alchemical simulations.

431e Exploration of the Retention Mechanism in Supercritical Fluid Chromatography by Molecular Simulations

Chun-Kai Chang, Joern Siepmann, Faith L. Wroniuk, Yih Ling Saw, James P. Grinias, Mark R. Schure, Stephanie A. Schuster

This presentation discusses molecular insights on the retention mechanism for prototypical analytes in supercritical fluid chromatography (SFC) and a comparison to the retention mechanism in reversed-phase liquid chromatography (RPLC). Gibbs ensemble Monte Carlo simulations are used to investigate the structure of the retentive phase in contact with different mobile phase compositions, temperatures, and pressures. The retentive phase model is a planar slit pore (width of 9 nm) with dimethyl octadecyl silane (ODS) ligands grafted to a cristobalite surface at a grafting density of 2.9 $\mu mol/m^2$ without end-capping that matches the one used in prior studies of RPLC retention [Rafferty *et al.*, *J. Chromatogr. A*, 1218 (2011) 2203]. The mobile phases investigated are CO_2/methanol mixtures (0, 5, and 10 mol% methanol in CO_2). The analyte molecules consist of C_1-C_6 linear alkanes and alcohols, and the conditions examined are at 293 (a subcritical temperature), 313, and 333 K with pressures of 15 and 30 MPa at each temperature. The molecular models are validated by comparison to experimental vaporliquid coexistence data for bulk CO_2/methanol mixtures.

The simulations indicate that the SFC retention of a methylene group (separating the homologous series of *n*-alkanes) is significantly weaker than in RPLC because the CO_2/methanol mixtures are much stronger solvents for the CH_2 group than neat water or an aqueous mobile phase with up to 50% v/v methanol or acetonitrile. In addition, the solubility of CO_2 in the ODS region is much larger than for water and methanol and makes the stationary phase less retentive for CH_2 groups. The presence of a hydroxyl group (separating a primary alcohol from the linear alkane with the same number of carbon atoms) reduces retention in RPLC because the mobile phase offers a more favorable environment than the stationary phase. The reverse is true in SFC with a neat CO_2 mobile phase where the availability of hydrogen-bonding sites due to residual silanols in the non-endcapped stationary phase makes it a more favorable environment. Addition of methanol allows for hydrogen-bonding in the SFC mobile phase and reduces the retention increment of the OH group due to stronger solvophilic interactions. Besides the stronger interactions with solvents, the methanol molecules "de-activate" the residual silanol groups, causing the loss of the retention of the OH group.

In addition to thermodynamic information, the simulation trajectories are analyzed to yield detailed information on the ODS conformations, the spatial distribution of CO_2 and methanol molecules in the stationary phase, the spatial and orientational preferences of analyte molecules, and the propensities for silanol-methanol, silanol-analyte, and methanol-analyte hydrogen bond formation. The chain conformation in the SFC system does not show substantial changes over the range of temperatures and pressures investigated here, whereas there are significant differences to the chain conformation in the RPLC systems.

This research is supported by the Chemical Measurement and Imaging program in the National Science Foundation Division of Chemistry under Grant No. CHE-2003246 (with partial co-funding from the Division of Chemical, Bioengineering, Environmental, and Transport Systems).

431f Impacts of Surface Functionalization on the Cellulose Nanocrystal Solubility in Ethanol

Arash Elahi, Santanu Chaudhuri

Cellulose-based materials have emerged as viable candidates for many industrial and academic studies due to their abundance, nontoxicity, and renewability. Recent progress in forming stable oil-in-water nanoemulsions using functionalized Cellulose Nanocrystals (CNC) yielded growing interest in adopting CNCs as emulsifiers. Nevertheless, the insolubility of CNCs in non-water solvents impeded them from stable nonaqueous emulsification, e.g., 2D material inks in ethanol.

Functionalization of CNCs' surfaces alters the hydrophilicity/hydrophobicity balance and potentially enhances their solubility in ethanol. Understanding the underlying physics of the agglomeration of CNCs will assist us in recognizing the influence of chemical modifications on CNCs' solubility in ethanol. We performed Molecular Dynamics simulations to evaluate the effects of surface modifications using several functional groups, e.g., alkyl, carboxyl, etc., on the behavior of Cellulose-ethanol systems. We used the extended version of the OPLS-AA force field for carbohydrates to model bond, angle, dihedral, and Van der Waals interactions.

We adopted the Alchemical free energy and Umbrella Sampling methods and investigated the impact of surface functionalization on Cellulose chain solubility and CNCs' dispersibility in ethanol. Further, we analyzed the variation of the tendency of CNCs to reaggregate as the function of the degree of substitution of the functional groups. Finally, we will explore the capability of the CNCs with improved solubility to exfoliate 2D materials, e.g., graphene, in ethanol.

431g Rational Design of Solvents for Catalytic Transformation of Biorenewable Platform Chemicals

Mohd Ussama, Gourav Shrivastav, M. Ali Haider

Valorizing biomass-derived lactones such as mevalonolactone (MVL), triacetic acid lactone (TAL), and γ-valerolactone (GVL) involve removal of oxygen-based functionalities to get a variety of high-value chemicals and liquid fuels. Reactions like dehydration, ring-opening, and decarboxylation reactions are often used to accomplish these conversions. Under specific catalytic and solvent conditions, these reactions may exhibit marked differences in reaction kinetics and selectivity. For example, under aqueous conditions, the ring-opening and decarboxylation reactions of TAL have been shown to occur at low temperatures with higher conversion and selectivity, demonstrating solvent's effect. Free energy calculations via ab-initio molecular dynamics (MD) simulations have also been performed to investigate the solvent's effect on saturated pyrones' ring-opening and decarboxylation reactions. Here, a similar study is conducted to elucidate the effect of polar (protic and aprotic) and apolar solvent media on the acid-catalyzed dehydration reaction of biomass-derived lactones. In further combination with classical MD simulations of bulk and mixed solvent environments, the solvent's spatial arrangement in the vicinity of reactant is studied.

Classical MD simulations are performed with GROMACS software package. Water and THF are used for polar and cyclohexane for solvent media. While the water is modeled using SPC/E potential, the reactant, 4-hydroxy-6-methyllactone (HML), THF, and cyclohexane are modeled using OPLS all-atom potential. The dehydration reaction of HML in the two solvent media is studied using Car-Parrinello Molecular Dynamics (CPMD) simulations. Ultrasoft psuedopotentials with BLYP functionals are used to treat the core and valence electrons, respectively. After initial wavefunction and geometry optimization, the simulation system are equilibrated for 50 ps. Subsequently, free energy calculations are performed in 2d collective variable space using metadynamics.

From both classical MD and CPMD, HML is observed to form 3 to 5 hydrogen bonds in water and 1 to 2 hydrogen bonds in THF (Fig.1(A)). For HML-H^+ complex in CPMD simulations, C-OH_2^+ bond length observed to be smaller in water than in THF for the reason of higher stabilization of the protonated hydroxyl group in water due to its high polarity. The effect of two solvents on the reaction energetics is evaluated using CPMD-metadynamics reactions. Mechanism for the dehydration of HML-H^+ is similar in both the solvents, that is, water removal, by the C-O bond cleavage, followed by the proton abstraction from β-carbon and formation of carbon-carbon double bond (Fig.1(B)). While in water, the average activation free energy barrier of TS is found to be around 133±8 kJ/mol, in THF, it is found to be ~107±9 KJ/mol (Fig.1(B)). The reduction of free energy barrier in THF depicts a relatively faster reaction in THF, and is found to be consistent with the experiments.

For the reaction in mixed solvent environment, we have first tried to understand the local solvation environment of the reactant using classical MD simulations for four different compositions of THF with mass fraction of 0.20, 0.40, 0.60 and 0.80 in water. Based on radial distribution functions for the oxygen atom of water in the vicinity of HML (center of mass), the water exhibits a higher tendency to solvate the hydroxyl group (Fig.1(C)). By motivating from this, we study the solvent effects on the acid catalyzed dehydration reaction on a broader level, protonated mevalonolactone (MVL) equilibrated in different solvents, water (polar protic), THF (polar aprotic) and cyclohexane (apolar) and in the mixture of these solvents, preliminary results showed the increase in water rich local domain near the reactant on increasing the organic phase.

Mechanistic insights are provided for the acid catalyzed dehydration of lactones in solvents of different nature and polarity, and their organization around the reactant can come up with an intimation to design the complex reaction environment for the conversion of biomass derived molecules.

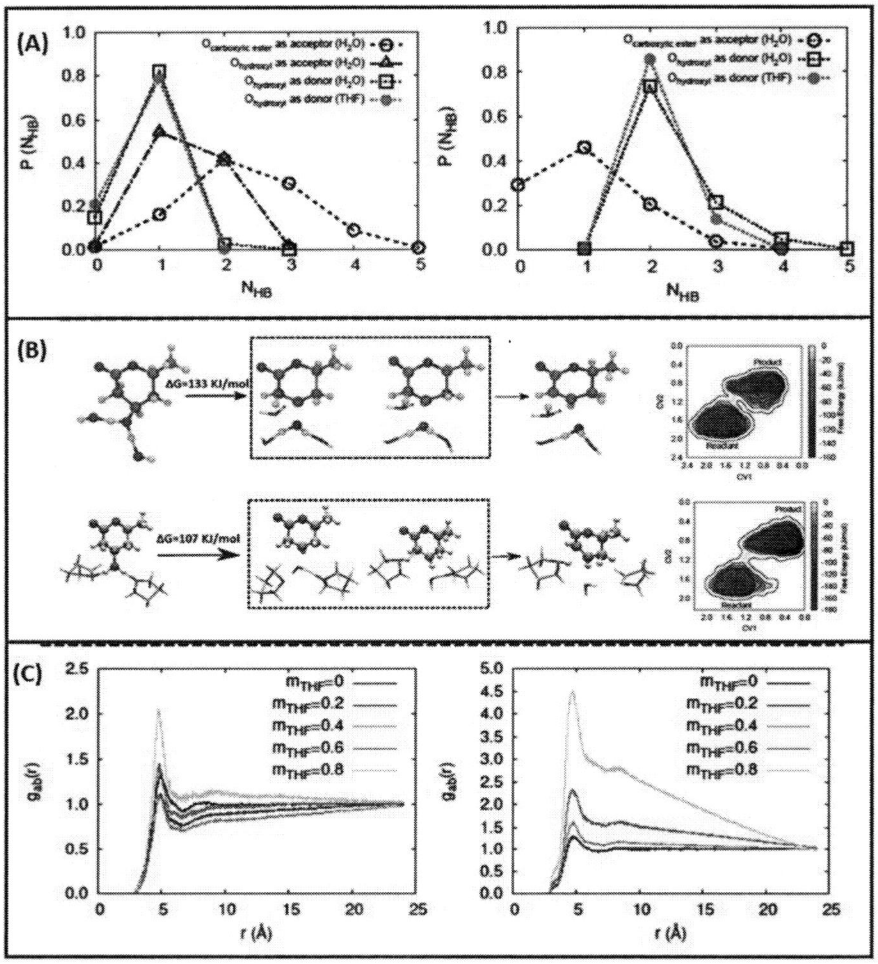

Fig.1 (A) Probability distribution P(NHB) of number of hydrogen-bonds in HML & HML-H$^+$ (B) Reaction and Gibbs free energy profile for HML dehydration in water & THF (C) RDF plot for different composition of water & THF

103a Brownian Bridges for Stochastic Chemical Processes – Applications and Approximation Method

Vivek Narsimhan, Shiyan Wang, Anirudh Venkatesh, Doraiswami Ramkrishna

A Brownian bridge is a continuous random walk conditioned to end in a given region of phase space. This phenomenon has many applications in chemical science where one wants to control the endpoint of a stochastic process – e.g., polymer physics, chemical reaction pathways, heat/mass transfer, and Brownian dynamics simulations. Despite its broad applicability, the biggest limitation of generating a Brownian bridge is that one needs to compute a drift that guides the random path to the end region with the correct statistics. This drift arises from the entropic penalty in constraining the random walk, and is derived from a Backwards Fokker Planck (BFP) equation that is infeasible to compute for complex systems. This talk introduces a fast approximation method to generate a Brownian bridge process without solving the BFP equation explicitly. Specifically, the talk will use the asymptotic properties of the BFP equation to generate an approximate drift, and determine ways to correct (i.e., re-weight) any errors incurred from this approximation. Because such a procedure avoids the solution of BFP equation, we show that it drastically accelerates the generation of conditioned random walks and allows the generation of such processes to be scaled to higher dimensions and complex systems. We will also show that this approach offers reasonable improvement compared to other importance sampling methods that use simple bias potentials.

103b Supervised Learning and the Finite-Temperature String Method for Computing Committor Functions and Reaction Rates

Muhammad Hasyim, Clay Batton, Kranthi K. Mandadapu

The discovery of transition pathways governing kinetic processes at the microscopic level is of fundamental interest in chemistry. Common to these problems is the presence of high-energy barriers, which make the pathway discovery difficult due to the need to sample rare events. The committor function, the probability that a trajectory starting from some initial configuration enters the product state before the reactant state, encodes the complete mechanistic details of these pathways, including the reaction rates and transition-state ensembles. Computing the committor function by traditional means is prohibitively expensive, requiring hundreds of trajectories per configuration to yield sufficient estimates along with the need for rare configurations along the transition pathway. Recent approaches have utilized the framework of transition path theory, in which the committor is a solution of a high-dimensional partial differential equation, to create data-driven approaches to estimate the committor function. Common to these approaches is the use of importance sampling, in which high probability regions of the transition pathway are targeted, and machine learning, in which an artificial neural network approximates the committor function. However, these prior approaches require either knowledge of suitable collective variables to perform metadynamics [1], or adaptively training the neural network via umbrella sampling [2], which can be demonstrated to not homogeneously sample the entire transition pathway without extensive fine-tuning.

Our work [3] extends the previous approach of [2] by utilizing supervised learning, in which sample-mean estimates of the committor function obtained via short simulation trajectories are used to fit the neural network, and the finite-temperature string method, a path-finding algorithm that enables homogeneous sampling across the transition pathway. We demonstrate these modifications on a two-dimensional Muller-Brown system as well as a model dimerization problem from molecular simulation, showing that they yield accurate estimates of the committor function and reaction rates. Additionally, an error analysis for our algorithm is developed from which the reaction rates can be accurately estimated via a small number of samples.

[1] Q. Li, B. Lin, and W. Ren. "Computing committor functions for the study of rare events using deep learning." The Journal of Chemical Physics 151, 054112 (2019).

[2] G. M. Rotskoff, A. R. Mitchell, and E. Vanden-Eijnden. "Active importance sampling for variational objectives dominated by rare events: Consequences for optimization and generalization." arXiv preprint arXiv:2008.06334v2 (2021).

[3] M. R. Hasyim, C. H. Batton, and K. K. Mandadapu. "Supervised Learning and the Finite-Temperature String Method for Computing Committor Functions and Reaction Rates." arXiv preprint arXiv:2107.13522 (2021).

103d Systematic Control of Collective Variables Learned from Variational Autoencoders

Jacob I. Monroe, Vincent K. Shen

Variational autoencoders (VAEs) are rapidly gaining popularity within molecular simulation for discovering low-dimensional, or latent, representations, which are critical for both analyzing and accelerating simulations. However, it remains unclear how the information a VAE learns is connected to its probabilistic structure, and, in turn, its loss function. Previous studies have focused on feature engineering, *ad hoc* modifications to loss functions, or adjustment of the prior to enforce desirable latent space properties. By applying priors of effectively arbitrary flexibility via normalizing flows, we focus instead on how adjusting the structure of the decoding model impacts the learned latent coordinate. We systematically adjust the power and complexity of the decoding distribution, observing that this has a significant impact on the structure of the latent space. By also varying weights on separate terms within each VAE loss function, we show that the level of detail encoded can be further tuned. We introduce a suite of broadly applicable metrics for generically assessing latent space structure and usage in order to rapidly characterize the utility of the learned collective variable without the need for in-depth prior knowledge of the system under study. By examining a wide variety of toy systems and molecules, we find that a meaningful latent space will only be learned if an appropriate balance is found between the complexity of the system, the power of the decoder, and the effective regularization strength. Our work provides practical guidance for utilizing VAEs to extract varying resolutions of low-dimensional information from molecular simulations.

103e Accelerated Free Energy Calculations By Joint Biasing in Configurational and Alchemical Space in Metadynamics

Wei-Tse Hsu, Pascal Merz, Giovanni Bussi, Michael Shirts

Alchemical transformation is a popular route adopted in free energy calculations associated with various biophysical processes, such as protein-ligand binding, or the transfer of a molecule from one environment to another. While the sampling of alchemical intermediate states in either parallel (e.g. Hamiltonian replica exchange) or serial manner (e.g. expanded ensemble) can bridge the gap in the configurational space between two end states of interest, relevant methods could fail in the scenarios where the most important slow degrees of freedom in the configurational space are almost orthogonal to the alchemical variable or the system gets trapped in a deep basin in the alchemical space.

To alleviate these issues, we propose to add alchemical variables as an allowed collective variable in metadynamics, augmenting the ability to sample collective variables. In this study, we validate our implementation of alchemical metadynamics in PLUMED with test systems/alchemical processes with varying complexities and dimensions of collective variable space, including the binding and unbinding of CB8-G3 host-guest binding complex. We show that one-dimensional alchemical metadynamics correctly recovers the free energy difference between the coupled state and the uncoupled state, and show how it can overcome free energy barriers orthogonal to alchemical directions. Multi-dimensional alchemical metadynamics has the potential to further facilitate the sampling by introducing configurational collective variables, now including alchemical parameters, to circumvent barriers in new ways. The method can trivially be combined with other metadynamics-based algorithms implemented in PLUMED. The necessary PLUMED code changes have already been released for general use in PLUMED 2.8.

103f Exploring Low-Energy Regions of Glassy Potential Energy Landscapes

Amruthesh Thirumalaiswamy, Robert Riggleman, John C. Crocker

Glassy materials are ubiquitous in human life, with applications ranging from composites in computing chips to dense emulsions in the everyday hand soap. The complex physics and dynamics of glassy materials has puzzled scientists for long. The inability of glasses to equilibrate and find their lowest energy configurations is often attributed to their potential energy landscapes: rugged, barrier-filled surfaces in high-dimensional space that seem incredibly difficult to navigate. Further, the characteristic properties of system configurations deep in the energy landscape are fundamental to understanding questions surrounding the glass transition. In this study, we employ a modified metadynamics based algorithm, and report the surprising observation of canyon-like structures in high-dimensional glassy energy landscapes. Our algorithm successfully uses these canyons as navigational aids to access low energy states deep within the potential energy landscape. The earlier-noted rugged surfaces line the canyon floors with smooth canyon walls. The algorithm successfully overcomes local ruggedness and explores low-energy regions of the landscape. Further, we observe such canyons in three vastly different glassy systems, successfully sampling low energy configurations in each - while revealing fractal signatures in the landscape structure. Finally, we contrast our results with ones obtained using conventionally studied techniques like molecular dynamics and Monte Carlo based schemes, discussing the pros and cons of each.

Funding source acknowledgement: NSF DMR-1609525

103g Ordinary Least Squares, Extraordinary Overconfidence: Towards Improved Statistical Estimators for Particle Diffusivity

Kevin S. Silmore, Yuanhao Li, Gerald J. Wang

Self-diffusivity and binary diffusivity are fundamental transport properties of interest in both molecular simulations and single-particle tracking experiments. Given an ensemble of particle trajectories, it is often the case in practice that an estimate of the diffusivity (and its associated uncertainty) is obtained by fitting a line to mean squared displacement data via ordinary least squares, based on the classic results of Einstein and Smoluchowski. In this work, we discuss why this procedure is statistically suboptimal and demonstrate how improved statistical procedures — both Bayesian and frequentist — can instead be applied to particle position data in order to estimate diffusivities with statistically rigorous uncertainties / error bars. Applications to simulations of bulk and nanoconfined fluids as well as nanoparticle single-particle tracking data will also be discussed.

103h How to Quantify Spatial Variations in Diffusivity in Molecular Dynamics Simulations.

Tiago Domingues, Ronald Coifman, Amir Haji-Akbari

The structure and function of many materials are strongly impacted by confinement as the properties of confined states of matter can deviate considerably from their bulk counterparts. At a fundamental level, confinement breaks the translational symmetry of materials, and makes all thermodynamic, structural and transport properties functions of position. Furthermore, transport properties usually become anisotropic (or direction-dependent) upon confinement. Characterizing such anisotropy and spatial variations is critical to understanding and engineering the behavior of materials whose properties are modulated by confinement. To this end, computational studies of confined materials have focused on ways of characterizing such spatial variations. This is a straightforward task for static properties, or mechanical observables that can be unambiguously calculated for each configuration, as their profiles can be readily computed via spatial binning of the simulation box. Such an approach cannot, however, be rigorously applied for dynamical properties such as transport coefficients as they can only be estimated from autocorrelations of mechanical observables, and there is no natural binning approach that can systematically capture spatial variations in such autocorrelations. For instance, diffusivity within bulk materials can be readily computed either from the mean-squared displacement (MSD) or using the velocity autocorrelation function (VACF). Both approaches break down, however, in the case of confined materials since no direct correspondence exists between any ad hoc notions of local MSD or VACF and the solutions of the Smoluchowski equation that governs anisotropic position-dependent diffusion.

In this work, we propose a generalized estimator for the position-dependent diffusivity tensor. We first validate this estimator by applying it to ensembles of trajectories generated via a stochastic differential equation (SDE) solver. We then apply this method to two systems, pure liquid within a slit pore and a supported film of a binary glass-forming liquid and obtain diffusivity profiles that yield autocorrelation functions that are consistent with those attained from MD simulations. Our approach can potentially provide a framework for characterizing position-dependent and anisotropic diffusivities in confocal microscopy experiments and molecular simulations.

103i Leveraging the Wolf Method for Electrostatics to Extend Time and Length Scales Accessible By Monte Carlo Simulations

Gregory Schwing, Mohammad Soroush Barhaghi, Brad Crawford, Loren Schwiebert, Jeffrey Potoff

Calculating the Coulombic potential through direct, pairwise r^{-1} summation is known as the Madelung problem. Exact solutions, such as Ewald summation, Particle Mesh Ewald (PME), and Particle-Particle Particle Mesh (P3M), require large time and memory usage.[1-3] In Monte Carlo simulations, it is common to use the Ewald summation method, although an approximation known as Wolf summation is faster and more memory-efficient.[4] The original Wolf method was derived for point-charge simulations, in recent years, three different formulations to handle intra-molecular interactions in molecular systems have been developed. These approaches include 1) Gross et al, (2) Vlugt et al, and (3) Vlugt with intra-cutoff (as implemented in the Cassandra Monte Carlo Engine).[5-7] Two potentials, Dampened Shifted Potential and Dampened Shifted Force were implemented for each method.[8] This work is the first comparative analysis of the three approaches. The accuracy and performance of the Wolf method, relative to the Ewald summation, is evaluated through the prediction of the vapor-liquid equilibria for water, ethanol free energy of hydration, and the hydration of complex biological structures.

For the models tested, there was no distinguishable difference in the single point electrostatic energies calculated with the Vlugt and Vlugt with intra-cutoff methods. The relative error of the Vlugt method converged to 0.0 as alpha approached 1.0, while the Gross implementation diverged. In grand canonical Monte Carlo simulations, simulations of liquid-liked densities produced density histograms in exact agreement with reference calculations performed with the Ewald summation. However, significant deviations between vapor-phase density distributions calculated with Wolf and Ewald methods were observed. Using GCMC to hydrate large biological systems, the density produced using the Wolf method was indistinguishable from Ewald simulations, while displaying over twice the computational efficiency. Simulations of a water cube with a box length of 1000 Angstrom required nearly 1 Terabyte of RAM when using the Ewald summation, while Wolf summation required less than 32 GB. Using the CPU version of GOMC[9,10] the Wolf summation was approximately 8x faster than the Ewald summation method.

References:

[1] Ewald, Paul P. 'Die Berechnung optischer und elektrostatischer Gitterpotentiale.' *Annalen der physik* 369.3 (1921): 253-287.

[2] Darden, T.; York, D.; Pedersen, L. Particle mesh Ewald: an N·log(N) method for Ewald sums in large systems. Journal of Chemical Physics 1993, 98, 10089–10092.

[3] Hockney, R. W.; Eastwood, J. W. Computer Simulation Using Particles; McGraw-Hill: New York, 1981.

[4] Wolf, D., et al. 'Exact method for the simulation of Coulombic systems by spherically truncated, pairwise r− 1 summation.' *The Journal of chemical physics* 110.17 (1999): 8254-8282.

[5] Waibel, Christian, and Joachim Gross. 'Modification of the Wolf method and evaluation for molecular simulation of vapor–liquid equilibria.' Journal of Chemical Theory and Computation 14.4 (2018): 2198-2206.

[6] Hens, Remco, and Thijs JH Vlugt. 'Molecular simulation of Vapor–Liquid Equilibria using the Wolf method for electrostatic interactions.' *Journal of Chemical andamp; Engineering Data* 63.4 (2017): 1096-1102.

[7] Cassandra: An Open Source Monte Carlo Package for Molecular Simulation', Jindal K. Shah, Eliseo Marin-Rimoldi, Ryan Gotchy Mullen,Brian P. Keene, Sandip Khan, Andrew S. Paluch, Neeraj Rai, Lucienne L. Romanielo, Thomas W. Rosch, Brian Yoo, and Edward J. Maginn, Journal of Computational Chemistry 2017, 38, 1727–1739

[8] Fennell, Christopher J., and J. Daniel Gezelter. 'Is the Ewald summation still necessary? Pairwise alternatives to the

accepted standard for long-range electrostatics.' *The Journal of chemical physics* 124.23 (2006): 234104.

[9] Y. Nejahi, M. Barhaghi, J. Mick, B. Jackman, K. Rushaidat, Y. Li, L. Schwiebert and J. Potoff, 'GOMC: GPU Optimized Monte Carlo for the simulation of phase equilibria and physical properties of complex fluids", SoftwareX, vol. 9, p. 20–27, 2019.

[10] Y. Nejahi, M. Barhaghi, G. Schwing, L. Schwiebert and J. Potoff, 'Update 2.70 to GOMC: GPU Optimized Monte Carlo for the simulation of phase equilibria and physical properties of complex fluids", SoftwareX, vol. 13, p. 100627, 2021.

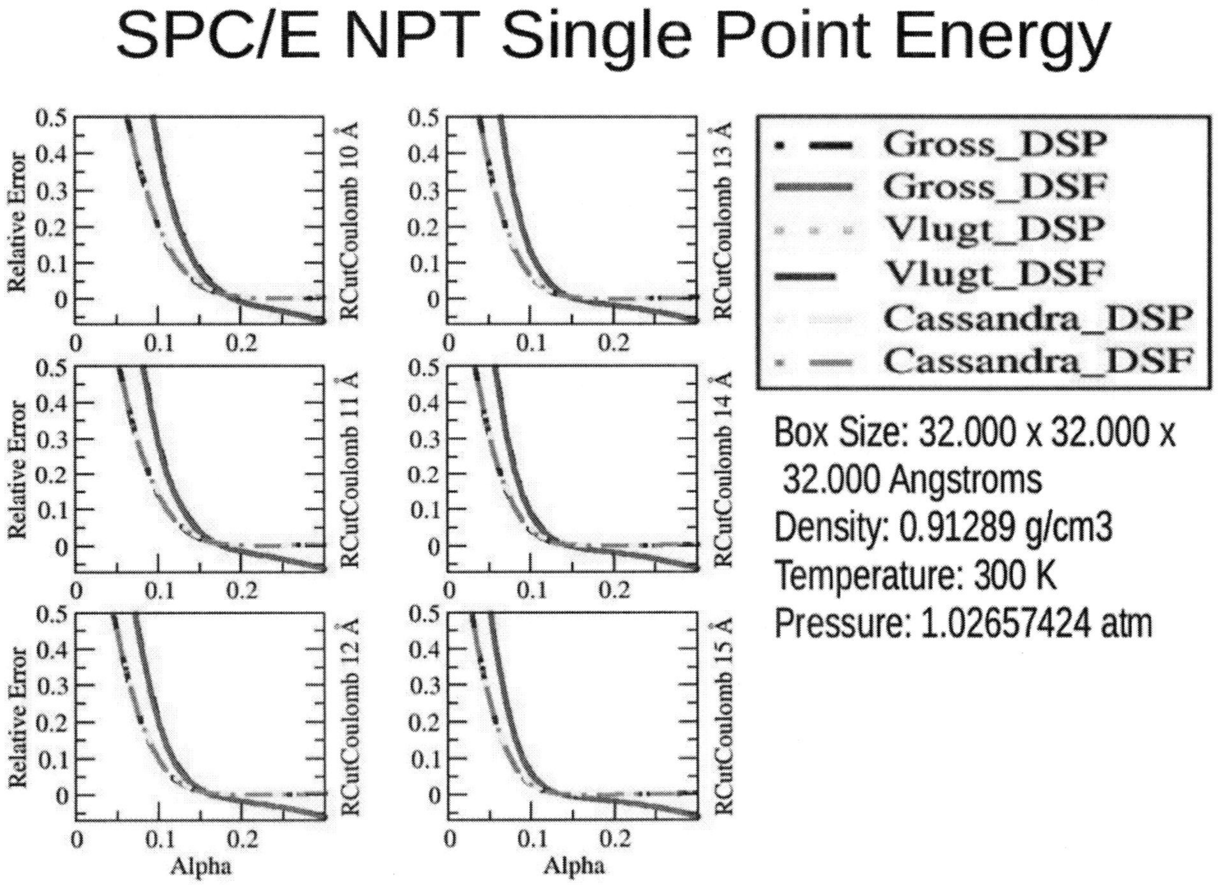

103j py-MCMD: Hybrid Monte Carlo – Molecular Dynamics Simulations

Brad Crawford, Mohammad Soroush Barhaghi, Gregory Schwing, David Hardy, John Stone, Loren Schwiebert, Jeffrey Potoff, Emad Tajkhorshid

Hybrid simulations utilize the unique strengths of Monte Carlo and molecular dynamics simulations to optimize the sampling of phase space, and enable simulations that are not possible with either simulation type on its own [1, 2, 3, 4]. In this work, we present the py-MCMD software, which allows Monte Carlo and molecular dynamics simulations to work together as a hybrid simulation. py-MCMD manages simulation execution and the data transfer between the simulation engines. GPU Optimized Monte Carlo (GOMC) [5, 6] is used for Monte Carlo moves, while NAMD [7, 8] is used for molecular dynamics sampling.

To validate the workflow, and highlight some of its capabilities, hybrid MC/MD simulations are performed for SPC/E water in the isobaric-isothermal (NPT), and grand canonical (GC) ensembles, as well as with Gibbs ensemble Monte Carlo (GEMC). The hybrid MC/MD approach agrees with reference MC simulations within the statistical uncertainty of the calculations, and has a computational efficiency that is 2 to 136 times greater than traditional Monte Carlo simulations. The influence of the integrator parameters, and parameters controlling the calculation of electrostatic energies, on the accuracy of the results, and the efficiency of the simulation, are discussed. MC/MD simulations performed for water in a graphene slit pore illustrate significant gains in sampling efficiency when using the coupled--decoupled configurational-bias (CD-CBMC) algorithm[9] compared to unbiased single random trial insertions. Simulations using CD-CBMC reach equilibrium 25 times faster than simulations using unbiased single random trial insertions. In a more challenging application, hybrid grand canonical Monte Carlo/molecular dynamics (GC MC/MD) simulations are used to hydrate a buried binding pocket in the bovine pancreatic trypsin inhibitor (BPTI). Water occupancies produced by GC MC/MD simulations are in close agreement with crystallographically identified positions. GC MC/MD simulations have a computational efficiency that is 5 times better than MD simulations. py-MCMD is available on GitHub at https://github.com/GOMC-WSU/py-MCMD.

References

[1] Y. Deng and B. Roux, 'Computation of binding free energy with molecular dynamics and grand canonical Monte Carlo simulations', J. Chem. Phys., vol. 128, p. 115103, 2008.

[2] G. Ross, E. Russell, Y. Deng, C. Lu, E. Harder, R. Abel and L. Wang, 'Enhancing Water Sampling in Free Energy Calculations with Grand Canonical Monte Carlo', Journal of Chemical Theory and Computation, vol. 16, p. 6061–6076, 2020.

[3] M. Samways, H. Bruce Macdonald and J. Essex, 'grand: A Python Module for Grand Canonical Water Sampling in OpenMM', Journal of Chemical Information and Modeling, vol. 60, p. 4436–4441, 2020.

[4] I. Ben-Shalom, C. Lin, T. Kurtzman, R. Walker and M. Gilson, 'Simulating Water Exchange to Buried Binding Sites', Journal of Chemical Theory and Computation, vol. 15, p. 2684–2691, 2019.

[5] Y. Nejahi, M. Barhaghi, J. Mick, B. Jackman, K. Rushaidat, Y. Li, L. Schwiebert and J. Potoff, 'GOMC: GPU Optimized Monte Carlo for the simulation of phase equilibria and physical properties of complex fluids', SoftwareX, vol. 9, p. 20–27, 2019.

[6] Y. Nejahi, M. Barhaghi, G. Schwing, L. Schwiebert and J. Potoff, 'Update 2.70 to GOMC: GPU Optimized Monte Carlo for the simulation of phase equilibria and physical properties of complex fluids', SoftwareX , vol. 13, p. 100627, 2021.

[7] J. Phillips, R. Braun, W. Wang, J. Gumbart, E. Tajkhorshid, E. Villa, C. Chipot, R. Skeel, L. Kale and K. Schulten, 'Scalable molecular dynamics with NAMD', Journal of Computational Chemistry, vol. 26, p. 1781–1802, 2005.

[8] J. Phillips, D. Hardy, J. Maia, J. Stone, J. Ribeiro, R. Bernardi, R. Buch, G. Fiorin, J. Hénin, W. Jiang, R. McGreevy, M. Melo, B. Radak, R. Skeel, A. Singharoy, Y. Wang, B. Roux, A. Aksimentiev, Z. Luthey-Schulten, L. Kalé, K. Schulten, C. Chipot, E. Tajkhorshid, ' Scalable molecular dynamics on CPU and GPU architectures with NAMD', Journal of Chemical Physics, vol. 153, p. 044130, 2020.

[9] Martin, M.; Siepmann, J., "Novel configurational-bias Monte Carlo method for branched molecules. Transferable potentials for phase equilibria. 2. United-atom description of branched alkanes', J. Phys. Chem. B vol. 103, p. 4508–4517, 1999.

233a Symmetric Molecular Dynamics

Andrew White

We derive a formulation of molecular dynamics that generates only symmetric configurations. We implement it for all 2D planar and 3D space groups. An atlas of 2D Lennard-Jones crystals under all planar groups is created with symmetric molecular dynamics.

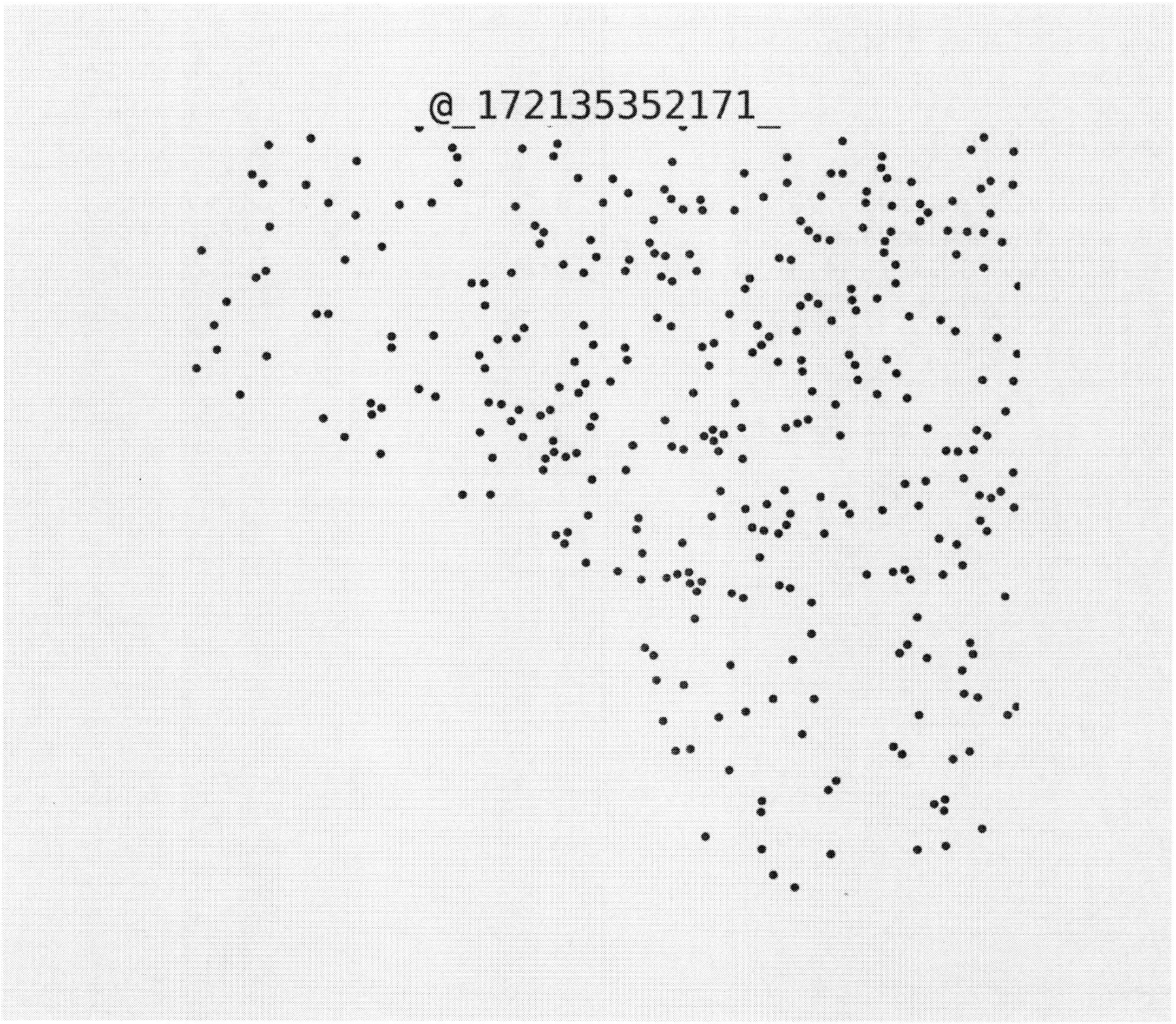

233b An Extended-Ensemble Relative Entropy Approach to Sequence-Specific Coarse-Grained Models for Peptide Aggregation

Evan Pretti, M. Scott Shell

Understanding the mechanisms behind aggregation of intrinsically disordered proteins is an important and practically relevant problem: for instance, tau protein aggregation is responsible for the class of neurodegenerative diseases known as tauopathies. However, aggregating systems pose unique challenges for molecular simulation; the ranges of length and time scales involved necessitate the use of multiscale modeling. Coarse-grained models capable of accurately predicting disordered protein aggregation, and its dependence on the protein sequence and aggregation conditions, must reproduce the relevant balance of interactions driving assembly. Empirically developed top-down models can reproduce features of an aggregation process at low computational cost. However, it can be difficult to use these models to understand how specific changes in the molecular details of systems can influence their behavior. Hybrid models incorporating information from both simulations and experimentally known structures can be useful if a previously characterized protein sequence is to be studied, but are less applicable for studying novel mutations. Purely bottom-up models, in principle, can be made completely flexible in this regard. These, however, must be able to capture the particular differences in interaction strengths between pairs of residues to allow modeled disordered proteins to sample the correct ensemble of disordered configurations as well as specific ordered aggregates.

Here, we present a novel approach to developing bottom-up coarse-grained protein models using relative entropy optimization. This "extended ensemble" method allows models to be optimized simultaneously against multiple atomistic reference systems, each containing multiple interacting polypeptide chains, over a range of temperatures. The interactions in the resulting models are thus transferable to new sequences. We demonstrate this method first on simplified reduced-alphabet systems containing representative amino acids, showing how the temperatures of the reference systems can be chosen to optimize the balance of disordered configurations and ordered secondary structures in the reference ensemble. The sequence-specific coarse-grained models made with this strategy accurately predict secondary structures as well as end-to-end distance distributions, even for sequences outside of the training set. Next, we apply the method to aggregation-prone fragments of tau protein that form parts of the ordered cores of many disease-associated fibrillar tau structures, specifically the PHF6 hexapeptide and the hairpin-forming HP1 19-mer. Models of these systems can assemble large numbers of monomers into paired helical aggregates known to be characteristic of experimentally observed fibrils. Overall, we demonstrate that this approach provides a robust, systematic bottom-up path to sequence-specific coarse-grained models useful for studying protein aggregation and assembly.

233c Drl Take the Wheel: Employing Deep Reinforcement Learning to Drive Vaccine Models

Jonathan Faris, Daniel Orbidan, Brenden Petersen, Kayla Sprenger

Highly mutable infectious disease pathogens (hm-IDPs) such as HIV and influenza evolve faster than the human immune system, allowing them to circumvent traditional vaccination approaches and cause over one million deaths annually. Agent-based models can be used to describe the complex interactions that occur between immune cells and hm-IDP-like proteins (antigens) during affinity maturation—the process by which antibodies evolve. As such, compared to existing experimental approaches, these models offer a safe, low-cost, and rapid route to study and assess the immune response to vaccines spanning a wide range of design variables. However, the highly stochastic nature of affinity maturation and vast sequence space of hm-IDPs render standard agent-based modeling approaches intractable for exploring all pertinent vaccine design variables and the subset of immunization protocols encompassed therein. Herein, we present a multi-scale model of the affinity maturation process—taking into account both sequence and structural level information—enabling the rapid screening of potential HIV-1 vaccination protocols (antigen design, concentration, number of inoculations). Further, we use a deep symbolic optimization framework to determine sequence-function determinants in the antigen sequence. Molecular dynamics simulations were then employed to develop an atomistic understanding of the mechanism by which the antibodies evolved are binding to the HIV Env protein. Future works include *in vitro* and *in vivo* studies to determine antibody stability and antigen immunogenicity.

233d Accelerated Kinetic Monte Carlo (kMC) Simulation and Density Functional Theory (DFT) to Predict Turn-over Frequencies in Heterogeneous Complex Catalytic Reactions

Silabrata Pahari, Chi-Ho Lee, Joseph Kwon

Density functional theory (DFT) involves the solution of the Schrodinger equation for the interacting electrons of a molecular structure. This method has extensively been explored to design novel catalysts. Additionally, this method has the capability to explore the electronic structure of system without introducing a large number of physical parameters. Therefore, many recent studies have investigated the catalytic performances of diverse materials using this method by considering the thermodynamics of the catalytic materials [1]. Using DFT analysis important structural features of the catalyst and electronic properties can be are derived. Therefore, these calculations help us obtain the energies associated with individual reaction pathways in heterogeneous catalysis [1,2]. However, ab-initio calculations performed with DFT fail to address issues related to the kinetics of the catalysts and therefore fail to provide us with insights into important catalyst properties such as adsorbent localization, surface coverage, and reaction turn-over frequencies (TOF) with varying pressure and temperature [2]. To this end, methods like kinetic Monte Carlo (kMC) simulation can be utilized [3,4]. However, the majority of complex heterogeneous catalysis reactions have pathways that involve elementary reactions occurring at disparate rates [5,6,7]. Since in the kMC algorithm, processes are selected based on their relative rates, it is much more likely that a faster process will occur at each step, with a very infrequent sampling of the slower processes [8]. The infrequent sampling of slower reaction steps significantly undermines the use of computational resources implemented to study the effect of slow but crucial reaction steps on the final catalytic activities [7]. Additionally, owing to a continuous sampling of fast processes, the kMC algorithm allocates most of its time in exploring kinetically trapped configurations of the system that provide no significant information regarding the evolution of the catalytic surface and makes it computationally intractable to reach properties like the TOF values or catalytic selectivity over longer time-scales [7,8]. Hence, in its conventional form, the kMC algorithm and DFT cannot be combined to complement each other in predicting important catalytic activities. Specifically, conventional kMC methods fails in cases where the DFT calculations yield elementary reaction steps with significantly different reaction rates.

Therefore, to apply the kMC algorithms alongside DFT calculations for a more general class of catalytic reactions it becomes important that efficient strategies be developed that allow us to effectively sample slow and fast catalytic reaction steps in a computationally tractable fashion [9]. To this end, we propose a novel method of combining DFT calculations with kMC simulations. Specifically, this begins by identifying the distinct reaction channels present in the reaction mechanism. In the proposed methods based on the products formed in the reactions the intermediates related to individual reaction channels are identified and segregated. Then by analyzing the activation energies obtained from DFT calculations for both the forward and reverse reactions a reaction channel is either denoted as reversible or irreversible. All the fast reactions in the catalytic surface are assumed to be reversible and accordingly the free energy of the reverse reactions is also obtained for individual reaction steps via DFT calculations. After the activation energies are obtained the rate expression using the Arrhenius law is derived for the individual reactions. Reaction rates in each of the reaction channels are adaptively coarse-grained by using a scaling factor which essentially increases the energy barriers of the fast reaction channels to bring the reactions rates of the fast reactions within a few orders of magnitude of the slow reactions. The reaction channels are coarse-grained when a channel is quasiequilbrated. The degree of quasiequilibration is determined by a threshold value which acts as a tuning parameter of the algorithm. As predicted by DFT calculations when the reaction channel with the highest energy barrier is executed the kMC method stops exploring a kinetic basin and hops to the adjacent kinetic basin and the scaling factors of all the reaction channels are reinitialized. This algorithm minimizes the re-sampling of similar system configurations and allows us to sample the slower reactions in a computationally efficient manner. As a case study, the accelerated kMC and DFT algorithm is implemented to study the complex nitrogen reduction reaction (NRR) reaction on heterogenous catalyst and obtain the TOF values.

References.

[1] Weijing, D., Weihong, Z., Xiaodong, Z., Baofeng, Z., Lei, C., Laizhi, S., ... andamp; Ge, S. (2018). The application of DFT in catalysis and adsorption reaction system. *Energy Procedia, 152*, 997-1002.

[2] Prats, H., Illas, F., andamp; Sayós, R. (2018). General concepts, assumptions, drawbacks, and misuses in kinetic M onte C arlo and microkinetic modeling simulations applied to computational heterogeneous catalysis. *International Journal of Quantum Chemistry, 118*(9), e25518.

[3] Reuter, K., Frenkel, D., andamp; Scheffler, M. (2004). The steady state of heterogeneous catalysis, studied by first-principles statistical mechanics. *Physical review letters, 93*(11), 116105.

[4] Stamatakis, M. (2014). Kinetic modelling of heterogeneous catalytic systems. *Journal of Physics: Condensed Matter, 27*(1), 013001.

[5] Chen, Z., Wang, H., Su, N. Q., Duan, S., Shen, T., andamp; Xu, X. (2018). Beyond mean-field microkinetics: toward accurate and efficient theoretical modeling in heterogeneous catalysis. *ACS Catalysis*, 5816-5826.

[6] Schulze, T. P. (2002). Kinetic Monte Carlo simulations with minimal searching. *Physical Review E, 65*(3), 036704..

[7] Chatterjee, A., andamp; Vlachos, D. G. (2006). Temporal acceleration of spatially distributed kinetic Monte Carlo simulations. *Journal of Computational Physics, 211*(2), 596-615.

[8] Chatterjee, A., andamp; Vlachos, D. G. (2007). An overview of spatial microscopic and accelerated kinetic Monte Carlo methods. *Journal of computer-aided materials design, 14*(2), 253-308.

[9] Snyder, M. A., Chatterjee, A., andamp; Vlachos, D. G. (2005). Net-event kinetic Monte Carlo for overcoming stiffness in spatially homogeneous and distributed systems. *Computers andamp; Chemical Engineering, 29*(4), 701-712.

233e Towards the Elimination of Backmapping in Multiscale Simulations

Nicholas Jackson

All-atom backmapping represents an important technique for bridging molecular simulations spanning distinct length scales. Despite the ubiquity of the procedure, it comes with many undesirable features including ad hoc one-to-many mappings, repeated thermodynamic resampling and minimizations, and complicated software workflows. Here, we introduce a general computational framework that leverages heteroscedastic gaussian processes within a deep kernel learning framework to demonstrate how fine-grained property distributions can be learned directly at the associated coarse-grained resolution, including all necessary configurational dependence required to reproduce the full finite-grained property distribution. By performing this coarse-graining, we show that all-atom backmapping techniques can be potentially avoided in multiscale simulation frameworks, significantly accelerating multiscale workflows. We demonstrate this approach within the context of Electronic Coarse-Graining in soft materials, where the all-atom, thermally averaged electronic density of states is learned directly at the coarse-grained resolution, eliminating any future recourse to all-atom resampling or ad nauseum electronic structure calculations.

233f Implicit Heuristic Model Captures Electrostatic Features of Cell Membrane Environment

Rituparna Samanta, Jeffrey J. Gray

Membrane proteins (MPs) play a critical role in our body, constituting about a third of all human protein. They are targets for more than half of all drugs. However, prediction and design of MPs are challenging due to the presence of diverse and complex lipid environment. Relative to soluble proteins, the model development of membrane proteins particularly lag behind due to the sparse and low-quality data, leading to overfit tools and specific models.

Implicit models accelerate this complex biomolecular problem by representing the solvent and the lipid environment as a continuum medium. Additionally, to overcome the challenge of sparse dataset, we assembled a suite of 12 tests on independent datasets ranging from predicting structural property, stability to protein-protein docking and design to test the model. Most implicit models often do not consider the effect of pH, lipid head group, or dielectric constant of membrane environment. In this work, we propose to develop an implicit approach that captures the crucial electrostatic interactions due to the membrane, such as the effect of lipid head groups the influence of pH and dielectric variations inside the membrane layer. Our energy function franklin2022 is built upon franklin2019, an existing energy function based on experimentally derived hydrophobicity scales that could capture the anisotropic structure, the shape of water-filled pores, and nano-scale dimensions of membranes with different lipid compositions. Our new method uses a constant-pH algorithm to sample the protonated and de-protonated states of protein residues. Further, it captures the effect of lipid head group using a mean-field based approach and uses a depth-dependent dielectric constant to characterize the membrane environment. Relative to franklin2019, this model has the ability to capture the effect of pH and improved the calculation of $\Delta\Delta G_{pH}$ of low pH insertion peptides (pHLIP) in extracellular acid environments, important biomarkers of cancer cells. Further, after including the effect of lipid head groups, franklin2022 have improved the prediction of tilt angles of adsorbed peptides relative to franklin2019. We will further test the performance of franklin2022 on the benchmark suite to evaluate its ability to predict the stability, structure, and design membrane proteins.

The speed of such implicit models and the model calibration based on diverse tests will help access biophysical phenomena at different time and length scales to accelerate the design pipeline for membrane proteins.

233g Hamiltonian-Resolution Replica Exchange Improves Flexible Backbone Protein Docking By Mimicking Induced-Fit Pathways

Ameya Harmalkar, Jeffrey J. Gray

Protein-protein interactions (PPIs) are involved in almost all biological processes in human health and disease. Understanding the structure of a protein complex can reveal biological mechanisms and suggest intervention strategies. Since experimental techniques are often infeasible, docking provides an alternative to elucidate structures and guide molecular engineering. Recently, AlphaFold *(Jumper et al. 2021; Evans et al. 2021)* and other deep learning algorithms *(Baek et al. 2021)* have made a breakthrough in the field of protein sequence-to-structure prediction. Yet, modeling protein interfaces and predicting conformational changes, remains an unsolved challenge, especially when there is significant conformational change in one or both binding partners. To overcome these limitations, we introduce a new enhanced sampling algorithm, called Hamiltonian-Resolution replica exchange (H-ResEx). The H-ResEx algorithm improves canonical sampling over rugged, atomistic energy landscapes by swapping conformations between the coarse-grained (CG) and all-atom (AA) states. To capture large-scale conformational changes, we mimic induced-fit approach of protein binding and sample backbone moves on-the-fly. We demonstrate the performance of our method for docking protein targets on a benchmark set of 100 proteins with moderate to high flexibility (unbound to bound RMSD over 1.2 Å up to 8 Å). Moreover, our advanced simulation approach equips the computational advantages of coarse-graining and requires 300-500 CPU hours for a docking simulation (depending on protein sizes), which is efficient compared to molecular dynamics-based approaches. With this work, we show that a CG/AA exchange scheme with conformational sampling shows substantial promise towards quantitative and qualitative modeling of protein complexes and their dynamic interactions. The proposed algorithm paves the way towards challenging applications in enhanced sampling of biomolecular systems, employing coarse-grained models for capturing conformations without compromising on the accuracy of the all-atomistic models.

233h Modeling the Influence of Glycosylation on Protein Interaction

Bradley Harris, Yihan Huang, Zachary Rollins, Karen A. McDonald, Somen Nandi, Priya Shah, Steven George, Roland Faller

Glycosylation is an important posttranslational modification of proteins and plays a crucial role in protein interactions. It also needs to be understood for optimization of recombinant protein expression. We study 3D structures of several glycoproteins in atomistic detail starting from known crystal structures to examine the effect of oligosaccharides on glycoprotein structure to evaluate the accessibility of the oligosaccharide chains to glycan modifying enzymes. Computational methods for modeling glycoprotein structures are employed to understand how changes in site-specific oligosaccharide composition and structure influence 3D glycoprotein structure, protein oligomerization, and product quality attributes such as activity/efficacy, stability, and immunogenicity.

As an example we study the interaction of the receptor binding domain of the Spike protein of SARS-CoV-2 with the ACE2 receptor which under glycosylation leads to stronger and longer ranged binding interactions between the proteins. Particularly, at shorter distances the interactions are between residues of the proteins themselves whereas at larger distances these interactions are mediated by the glycans. We see a catch slip type behavior where interactions during pulling break and are taken over by new interactions forming. The dominant interaction mode are hydrogen bonds, but Lennard-Jones and electrostatic interactions are relevant as well.

T validate our methodology we also study a fully glycosylated computational model of the T Cell Receptor (TCR) bound to the peptide-Major Histocompatibility Complex. We find that weak glycan-protein or glycan-glycan interactions impact the equilibrated structure of the TCR and pMHC leading to an increase in the overall bond strength of the TCR-pMHC complex including the duration and energetic strength under constant load.

We generally determine that computational modeling can elucidate the importance of glycosylation in the context of protein interaction

233i Multiscale Molecular Modeling of Flory-Huggins χ-Parameter Calculation and Simulation of Structural Transformation of Photo-Regulated Multicompartment Micelle

Jinwon Cho, Ji Il Choi, Seung Soon Jang

There has been a growing interest in designing and synthesizing polymer nanostructures with degrees of complexity and controlling over the composition for various applications, which includes diagnostic imaging, and gene and drug delivery. Although much progress has been made in drug delivery systems, the design of a suitable carrier for the delivery of hydrophobic drugs is still a challenging problem. Overcoming such problem in designing drug carrier, multicompartment polymeric micelles, comprised of block copolymers with two or more monomeric species, have emerged as a solution since a desirable micellar architecture can be achieved by tuning polymer properties. Due to the complexity of the multicompartment micelle structure, molecular modeling and simulation have been widely used as powerful tools to characterize the structures and behaviors of micelles consisting of many amphiphilic multiblock copolymers in solution state. In this regard, the structural transformation of multicompartment micelle consisting of amphiphilic *A*-*b*-(***C-spiropyran***) diblock copolymer in water, induced by the photo-regulated transition between hydrophobic non-ionic ***spiropyran*** and hydrophilic zwitterionic ***merocyanine***, is investigated using a set of computational methods such as density functional theory (DFT), molecular dynamics (MD) simulation, and dissipated particle dynamics (DPD) simulation. By employing a new computational framework for Flory-Huggins -parameter, which is improved by considering solvation free energy, it is demonstrated that the -parameter for a ***merocyanine***-water pair is significantly smaller than that for a ***spiropyran***-water pair, indicating that hydrophobic ***spiropyran*** becomes hydrophilic ***merocyanine*** through the photo-switched transition. Our computational procedure of -parameter calculation is further validated by investigating the deprotonation of acrylic acid in water. Finally, using a set of calculated -parameters, DPD simulations are performed employing two different block copolymers with two different block length: A_{10}-*b*-(C_6-***Spiro***$_{12}$) (***Micelle I***) and A_6-*b*-(C_{10}-***Spiro***$_{12}$) (***Micelle II***), demonstrating that both micelles undergo distinctive inside-out transformations as a function of the block length ratio. Through the structure analysis, it turns out that ***Micelle I*** keeps a well-defined core-shell structural feature before and after the transformation, whereas ***Micelle II*** loses such structural feature because short ***Block A*** is not capable of forming the core of micelle.

687a VACUUMMS: stewardship of research software from the void

Frank Willmore

Software and software components developed in the course of graduate work are an important research product and their stewardship is a critical part of the research enterprise, both for the reproducibility of published work and for their application to new work. A narrative is presented for a set of analytics which was developed for the characterization of free space in glass-forming materials, and was lost to the research community for some time. These codes were ultimately re-packaged using CMake and spack, and released as VACUUMMS (Void Analysis Codes and Unix-like Utilities for Molecular Modeling and Simulation) and are now finding new life.

687b Polymerizeit! Software to Generate Molecular Structures of Crosslinked Polymers in silico

Salman Bin Kashif, Sapna Sarupria

Sarupria group has also developed PolymerizeIt. This software provides a modular, extensible, Python-based platform to generate structures of crosslinked polymers for molecular simulations. Most existing software are either in languages not as widely used as Python or have the crosslinking protocol hardcoded. On the other hand, PolymerizeIt has been designed to enable extensions and community contributions. In my talk, I will give an overview of PolymerizeIt. I will discuss the lessons we have learned and the challenges we have encountered in developing PolymerizeIt and plans for keeping the software sustainable.

687c Validating a New Force Field and Extensible Analysis for a Large-Scale Screening Study of P3HT

Jenny Fothergill, Eric Jankowski

Validating computational studies is challenging because (a) identical reproduction of a previously-used software and hardware may not be possible, (b) assumptions used during modeling and analysis may not be documented or clearly communicated, and (c) the computational effort for doing so may be prohibitive. Here we demonstrate an example validation study of the organic semiconducting polymer P3HT enabled by MoSDeF tools. We specify the software stack using docker containers and conda environment files. We document model, simulation input, and analysis details using the MoSDeF tools Foyer, mBuild, and Freud. We define a simulation state space using signac, and use signac-flow to manage submission of jobs to HPC clusters. We perform molecular dynamics simulations using HOOMD-Blue replicating the state space sampled in prior P3HT work with ~400 GPU-hours of computational effort, using GAFF-UA in place of the OPLS-UA force field. We validate the use of GAFF-UA in this work, observing that order-disorder transition temperature and scattering patterns follow the same dependence on solvent quality seen with OPLS-UA. We conclude with a discussion of additional work making structural analysis more transparent, reproducible, usable by others, and extensible (TRUE) with the GRiTS package we develop.

687d MoSDeF-GOMC: Python software for the creation of scientific workflows for the Monte Carlo simulation engine GOMC

Brad Crawford, Umesh Timalsina, Co Quach, Nicholas C. Craven, Justin Gilmer, Peter Cummings, Jeffrey Potoff

Monte Carlo and molecular dynamics simulations rely on several input files that include information such as atomic coordinates, molecular topology, force field parameters, and a control file that directs the behavior of the simulation engine. Traditionally, these files were generated by hand by users with expert knowledge. The need for expert knowledge makes the learning curve steep for new users, while the bespoke generation of input files is error-prone with poor reproducibility.

In this work, we present updates to MoSDeF-GOMC, a python interface to the Molecular Simulation Design Framework (MoSDeF) [1-5] that enables users to create all the input files required to perform simulations with the GPU Optimized Monte Carlo (GOMC) simulation engine [6-7]. MoSDeF-GOMC dramatically simplifies the process of building systems and assigning force field parameters. Additionally, it provides some expert-system features, guiding users towards reasonable values for numerous parameters used to control Monte Carlo simulations. When combined with the Signac software [8-9], complex workflows may be created that incorporate thousands of discrete simulations, supporting the use of MoSDeF and GOMC for high-throughput screen applications. To highlight some of the capabilities of MoSDeF-GOMC, a number of illustrative applications are presented, including the prediction of the vapor-liquid coexistence curve for the jet fuel surrogate S-8, the hydration free energy for noble gases in water, the adsorption of ethane in a metal organic framework, and gas adsorption in a polymer matrix.

References

[1] Y. Nejahi, M. Barhaghi, J. Mick, B. Jackman, K. Rushaidat, Y. Li, L. Schwiebert and J. Potoff, 'GOMC: GPU Optimized Monte Carlo for the simulation of phase equilibria and physical properties of complex fluids", SoftwareX, vol. 9, p. 20–27, 2019.

[2] Y. Nejahi, M. Barhaghi, G. Schwing, L. Schwiebert and J. Potoff, 'Update 2.70 to GOMC: GPU Optimized Monte Carlo for the simulation of phase equilibria and physical properties of complex fluids", SoftwareX, vol. 13, p. 100627, 2021.

[3] A. Summers, J. Gilmer, C. Iacovella, P. Cummings and C. McCabe, 'MoSDeF, a Python Framework Enabling Large-Scale Computational Screening of Soft Matter: Application to Chemistry-Property Relationships in Lubricating Monolayer Films", J. Chem. Theory. Comput., vol. 16, no. 3, p. 1779-1793, 2020.

[4] M. Thompson, R. Matsumoto, R. Sacci, N. Sanders and P. Cummings, 'Scalable Screening of Soft Matter: A Case Study of Mixtures of Ionic Liquids and Organic Solvents", J. Phys. Chem. B, vol. 123, no. 6, p. 1340–1347, 2019.

[5] MoSDeF - the Molecular Simulation Design Framework, 'https://github.com/mosdef-hub", 2019. [Online]. [Accessed March 2021].

[6] C. Klein, A. Summers, M. Thompson, J. Gilmer, C. McCabe, P. Cummings, J. Sallai and C. Iacovella, 'Formalizing atom-typing and the dissemination of force fields with foyer", Computational

Materials Science, vol. 167, p. 215-227, 2019.

[7] M. Thompson, J. Gilmer, R. Matsumoto, C. Quach, P. Shamaprasad and A. Yang, 'Towards molecular simulations that are transparent, reproducible, usable by others, and extensible (TRUE) ", Mol. Phys., vol. 118, p. e1742938, 2020.

[8] V. Ramasubramani, C. Adorf, P. Dodd, B. Dice and S. Glotzer, 'signac: A Python framework for data and workflow managemen", in Proceedings of the 17th Python in Science Conference, 152-159, 2018.

[9] C. Adorf, P. Dodd, V. Ramasubramani and S. Glotzer, 'Simple data and workflow management with the signac framework", Computational Materials Science , vol. 146, no. C, p. 220-229, 2018.

687e Chemprop: Machine Learning for Molecular Property Prediction

Charles McGill, Esther Heid, Yunsie Chung, Kevin Greenman, David Graff, Mengjie Liu, Camille Bilodeau, Rafael Gomez-Bombarelli, Connor Coley, Klavs Jensen, Tommi S. Jaakkola, Regina Barzilay, William Green

Chemprop is an easy-to-use machine learning software for chemical property prediction. The software is open-source and available for public use [1] . With a simple interface and contained workflow, Chemprop is an accessible tool for use by scientists with varying levels of engagement with machine-learning techniques who want to create high-quality models from their data. Supported by the Machine Learning for Pharmaceutical Discovery and Synthesis Consortium [2], the software has an active base of industry users, driving continuous development of new features and support.

We will discuss the architecture used by Chemprop, some notable examples of recent applications, and several of the software's significant features. Inputs to Chemprop models are provided as SMILES strings from which the software can construct 2D connectivity graphs of molecules. Chemprop's network achitecture uses directed message passing neural networks (d-MPNN) for learnable molecular encodings, as implemented and benchmarked against other architectures by Yang et al. [3]. The end-to-end learning enabled by this architecture allows for the software to extract the information from the molecular graph that is most relevant to the property target being modeled. The software has been used for a variety of different applications, showing the versatility of learned encodings: enthalpy of formation, activation energy, solubility, antibiotic activity, reaction regioselectivity, UV-Vis absorption, and infrared spectra.

The implementation of Chemprop has incorporated a number of features and functions to fit the needs of its users. Chemprop has GPU-enabled training and prediction of models. Additional functions have been added for hyperparameter optimization, the extraction of molecule latent representations, and the estimation and calibration of model uncertainty. Additional molecule- and atom-level features can be provided to bring in outside information from outside methods such as experimental measurements or quantum mechanical calculations. Chemprop supports inputs using reactions or multiple molecules (e.g., solvent and solute). Tools for transfer learning and weighted multitask models enable the model to infer useful relationships across distinct datasets. The contained workflow allows for users to perform all of Chemprop's main functions with minimal coding required.

[1] Chemprop: Molecular Property Prediction. https://github.com/chemprop/chemprop

[2] Machine Learning for Pharmaceutical Discovery and Sythesis Consortium. https://mlpds.mit.edu

[3] Yang, K.; Swanson, K.; Jin, W.; Coley, C.; Eiden, P.; Gao, H.; Guzman-Perez, A.; Hopper, T.; Kelley, B.; Mathea, M.; et al. Analyzing Learned Molecular Representations for Property Prediction. J. Chem. Inf. Model. 2019, 59 (8), 3370–3388. https://doi.org/10.1021/acs.jcim.9b00237.

687f Workflow Automation in Predicting Exciplex Formation in Arene-Amine Complexes

Abhilash Patra, Shaama Mallikarjun Sharada

The formation of excited-state complexes or exciplexes in specific acceptor-donor systems leads to prominent red-shifted emission peaks in the fluorescence spectra. In particular, the photon absorbed excited state of oligo-(p-phenylene) (OPP*) forms the exciplex with the ground state triethylamine (TEA)[1]. The red-shifted emission peak is broader than that of isolated OPP*, indicating that the exciplex is likely comprised of an ensemble of structures. Identification of even a single exciplex geometry requires significant computational effort towards geometry optimization with time-dependent density functional theory (TDDFT) and user-intervention to ensure that the optimization always follows the correct excited state. To address this problem for the ensemble of structures, we construct automated workflows for TDDFT optimization as implemented in Q-Chem [2]. Our workflow takes advantage of FireWorks[3] for job handling, Pymatgen[4] for input/output parsing, and Custodian[5] for error handling, in addition to our in-house excited-state properties parsing code. The exciplexes are verified by new emission bands at longer wavelengths, otherwise absent in the spectra of isolated molecules. We also analyze the effect of solvent dielectrics on emission bands. The developed workflow can be employed to study the emission intensity variation with the electron-donor concentration.

References:

1. K. J. Kron et al, *The Journal of Physical Chemistry A (page and volume not assigned yet) (2022).*

2. *Y. Shao et al., Molecular Physics, **113**(2), 184-215 (2015).*

3. *A. Jain et al., Concurrency and Computation: Practice and Experience, **27**(17), 5037-5059 (2015).*

4. *S. P. Ong et al., Computational Materials Science, **68**, 314–319 (2013).*

5. *K. Mathew et al., Computational Materials Science, **139**, 140-152 (2017).*

687g Parallelization of Grand Canonical Ensemble Monte Carlo Using Prefetching and Windowing of Flat Histogram Simulations

Harold Hatch, Vincent K. Shen

While simulation techniques such as Molecular Dynamics (MD) efficiently utilize multi-core processors using domain decomposition, Monte Carlo (MC) simulations with single particle perturbations are difficult to domain decompose especially when interaction ranges are not significantly shorter than the domain. Thus, although serial MC simulations may be more efficient than serial MD for computing phase equilibrium properties, for example, parallel MD is often used when parallel MC software and algorithms are less abundant. However, parallel prefetching Monte Carlo simulations in the canonical ensemble (CE) can speed up a simulation by a factor of 3 when using 4 cores [1]. In this work, prefetching in the grand canonical ensemble (GCE) is shown to be about as efficient than in the CE, despite counteracting non-trivial efficiency differences in acceptance probability and processor load balancing. In addition, serial efficiency is also improved by a systematic study of simulation parameters. Finally, in addition to parallelization via prefetching, parallelization efficiency in transition-matrix MC simulations is also investigated by decomposing a range of macrostates into windows. These parallelization techniques, made available in the open-source software FEASST, reduce the time required to compute a vapor-liquid phase diagram of Lennard-Jones or SPC/E water by over an order of magnitude.

[1] H. W. Hatch, "Parallel Prefetching for Canonical Ensemble Monte Carlo Simulations", J. Phys. Chem. B, 2020, 124, 7191-7198.

687h Pscf+: An Extended and Improved Open-Source Software Package for Polymer Self-Consistent Field Calculations of Block Copolymer Self-Assembly

Juntong He, Qiang Wang

We have extended the newly released C++/GPU version of PSCF[1], an open-source software package for real-space self-consistent field (SCF) calculations of the "standard" model (*i.e.*, incompressible melts of continuous Gaussian chains with the Dirac δ-function potential) for block copolymer self-assembly, to include the dissipative particle dynamics (DPD) model (*i.e.*, compressible melts of discrete Gaussian chains with the DPD potential), and used it to study the stability of various Frank-Kasper phases formed by neat diblock copolymer melts. SCF calculations of the DPD model typically having 10 segments on each diblock copolymer chain are at least one order of magnitude faster than, and give qualitatively different phase diagrams from, those of the "standard" model. We have also improved several aspects of PSCF[1], making it much more suitable for efficient construction of accurate phase diagrams for block copolymer self-assembly.

1. Cheong, G. K.; Chawla, A.; Morse, D. C.; Dorfman, K. D., Open-source code for self-consistent field theory calculations of block polymer phase behavior on graphics processing units. *Eur. Phys. J. E* **2020,***43*(2), 15.

297a Embedding Stereoelectronic Effects into Molecular Representations for Machine Learning

Gabe Gomes

Molecular representation learning lies at the intersection of machine learning and chemistry, being crucial for property prediction tasks. To date, many representation methods have been developed, ranging from simple physicochemical descriptors to sophisticated geometrical modeling. However, existing approaches typically ignore non-bonding electronic information, and are limited to capturing covalent interactions. To address this problem, we propose a novel molecular representation that generalizes to other interactions by incorporating stereoelectronic effects. We accomplished this goal by taking advantage of a well-known method that captures complex electronic interactions, Natural Bond Orbital (NBO) analysis. To limit the scope of this work to organic molecules, we applied NBO analyses for the structures from the QM9 dataset. Subsequently, we created a graph neural network method capable of predicting these interactions in a supervised learning fashion. The method yields a heterogeneous graph with representations for atom, bond, and lone pair orbitals while adding atom-specific orbitals and second-order interaction information. Our method also suggests a more general way to represent molecular structures, which supplements the use of nodes for atoms and edges for chemical bonds with the addition of bond orbital nodes and second-order interaction edges. We developed an end-to-end pipeline that takes a molecular geometry input and returns a representation with stereoelectronic information. The first step was to enrich the standard molecular graphs with lone pairs (predicted by a separate network) and bond orbital nodes, which are solved analytically for a given geometry. Second, we developed a multi-task approach that learns all the interaction values provided by the NBO analysis. Finally, we use the resulting representations with infused stereoelectronic information for downstream tasks, with experiments showing the approach's applicability to increase performance. Our approach is uniquely suited for optimal learning with limited datasets, with strong applicability to catalysis and reaction design.

297b Predicting Solubility Limits of Organic Solutes for a Wide Range of Solvents and Temperatures Using Machine Learning and Thermodynamics

Florence Vermeire, Yunsie Chung, William Green

The solubility of organic molecules is crucial in organic synthesis and industrial chemistry, it is important in the design of many phase separation and purification units, and it controls the migration of many species into the environment. Despite its importance, no tool exists to predict the solubility of solids in different organic solvents at various temperatures. Here we present a fast and convenient computational method for estimating the solubility of solid neutral organic molecules in water and many organic solvents for a broad range of temperatures. We provide a method, software package, and on-line tool for the prediction of the solubility based on only the molecule identifiers (SMILES or InChI) of the solute and solvent.

Specific to the pharmaceutical industry, the solubility of active pharmaceutical ingredients (API) in a variety of organic solvents is an important property in the development of new drugs. Many curated databases are available for the aqueous solid solubility of API's as this property is of interest in the initial screening process of potential drugs. Further in the drug development chain during lab-scale synthesis, purification, crystallization, and scale-up from batch to continuous processes, information on the solubility of the API's in organic solvents other than water is required. Despite of the tremendous progress made in machine learning for molecular and material science, for the direct prediction of solid solubility in organic solvents at different temperatures, the application of data driven methods is limited by data abundance. For this reason, data-driven machine learning methods are combined with computational chemistry and thermodynamics for the accurate prediction of solid solubility.

The method uses machine learning for the fast prediction of important properties and thermodynamics to relate those properties to the solid solubility. The machine learning models are used to predict solvation free energy[1], solvation enthalpy, Abraham solute parameters[2], and aqueous solid solubility at 298K. A transfer learning method combines quantum chemistry and experimental data for the prediction of solvation free energy and enthalpy to broaden the application range of the models to solutes with a higher molar mass[1]. The Abraham solute parameters are used to calculate the sublimation enthalpy, and the gas and solid phase heat capacities at 298K with correlations recently published by Abraham and Acree[3, 4]. For temperatures up to 350K, no empirical data is required for the prediction of solid solubility limits. At higher temperatures, the temperature dependence of the dissolution enthalpy has to be accounted for through the temperature dependence of the solvation free energy. To calculate the solvation free energy at higher temperatures we rely on our previously published method[5], for which the solvent's critical properties are required.

To train the machine learning submodels and to assess the accuracy and robustness of the new method, several quantum chemical and experimental datasets were constructed or compiled in this work. They are all collected in the SolProp data collection, which is made publicly available on Zenodo (https://zenodo.org/record/5970538). In the new CombiSolu-Exp and CombiSolu-HighT-Exp databases, experimental solid solubility data in pure organic solvents at different temperatures are collected from over 170 different literature sources. They are used to validate the new method for predicting solid solubility limits for more than 6000 datapoints including more than 115 solutes, 95 organic solvents, and temperatures up to 593K. Data from different sources are compared to assess the experimental uncertainty in published solid solubility data. For aqueous solid solubility, the mean-absolute-error (MAE) between different data sources goes up to 0.17 log10(mol/L), while for the solid solubility in organic solvents more than 12% of the duplicated data has an absolute deviation higher than 0.2 log10(mol/L). It should be noted that this rather high experimental uncertainty can significantly affect the perceived performance of the new method and machine learning models.

First, the solid solubility at 298K in different organic solvents is predicted by the proposed model. For 1051 datapoints in the CombiSolu-Exp database (98 solutes and 87 solvents), using only molecular identifiers of the solute and solvent, the solid solubility is predicted with a MAE equal to 0.62 log10(mol/L). This method uses machine learning models to predict the aqueous solid solubility, which is used as a reference to relate the solubility in water to the solubility in other solvents through thermodynamic relationships. Using the experimental solid solubility in ethanol as a reference improves the predictions to a MAE of 0.16 log10(mol/L). However, the experimental measurements in ethanol often originate from the same source as the experimental data measured in other solvents for the same solute. As a result, it is expected that some consistent errors specific to the experimental apparatus or procedure are compensated for and the model performance might be overestimated. Alternatively, we used the experimental aqueous solid solubility from different sources as a reference which led to a MAE of 0.35 log10(mol/L) for the prediction of solid solubility in different organic solvents at 298K.

To relate the solid solubility at 298K with the solid solubility at different temperatures, the model uses the dissolution enthalpy. This value is calculated through thermodynamic relations with the solvation enthalpy and sublimation enthalpy at 298K. Up to 350K, the temperature dependence of the dissolution enthalpy can be neglected and the solvation enthalpy at 298K, predicted by a transfer learning model, can be used for its calculation. The 4922 solid solubility datapoints in the CombiSolu-Exp database (115 solutes and 95 solvents between 243 and 364K) are predicted using only molecular identifiers of the solute and solvent with a MAE equal to 0.99 log10(mol/L). Using experimental solid solubility measured in ethanol as a reference - if available - improves the MAE to 0.29 log10(mol/L). At higher temperatures (>350K), the temperature dependence of the dissolution enthalpy has to be accounted for. This value can be calculated using the temperature dependent solvation enthalpy and the sublimation enthalpy at 298K. The temperature dependent solvation enthalpy can be calculated with our previously published method[5] and the solvent's critical properties. This method is validated with experimental solid solubility data from the CombiSolu-HighT-Exp database for 1306 datapoints, including 67 solutes and 15 solvents. The solubility trend as a function of temperature is captured well for temperatures up to 593K.

The developed model has an excellent performance in predicting solubility trends in different solvents at 298K and as a function of temperature. The performance can be seen in the attached figure for different solutes. Even though the absolute solid solubility prediction can be off in some cases, the ability to predict solubility trends has tremendous applications. With additional user information on the solubility of the solute in one organic solvent at room temperature (298K), the accuracy of the method can be improved. The newly developed method can be accessed through our conda package (https://anaconda.org/fhvermei/solprop_ml) and through our user-friendly web interface (https://rmg.mit.edu/database/solvation/searchSolubility/).

[1] Vermeire, F. H.; Green, W. H., *Transfer learning for solvation free energies: From quantum chemistry to experiments*. Chemical Engineering Journal **2021,** 418, 129307.

[2] Chung, Y.; Vermeire, F. H., et al., *Group Contribution and Machine Learning Approaches to Predict Abraham Solute Parameters, Solvation Free Energy, and Solvation Enthalpy*. Journal of Chemical Information and Modeling **2022,** 62, (3), 433-446.

[3] Abraham, M. H.; Acree, W. E., *Estimation of heat capacities of gases, liquids and solids, and heat capacities of vaporization and of sublimation of organic chemicals at 298.15 K*. Journal of Molecular Liquids **2020,** 317, 113969.

[4] Abraham, M. H.; Acree, W. E., *Estimation of enthalpies of sublimation of organic, organometallic and inorganic compounds*. Fluid Phase Equilibria **2020,** 515, 112575.

[5] Chung, Y.; Gillis, R. J., et al., *Temperature-dependent vapor–liquid equilibria and solvation free energy estimation from minimal data*. AIChE Journal **2020,** 66, (6), e16976.

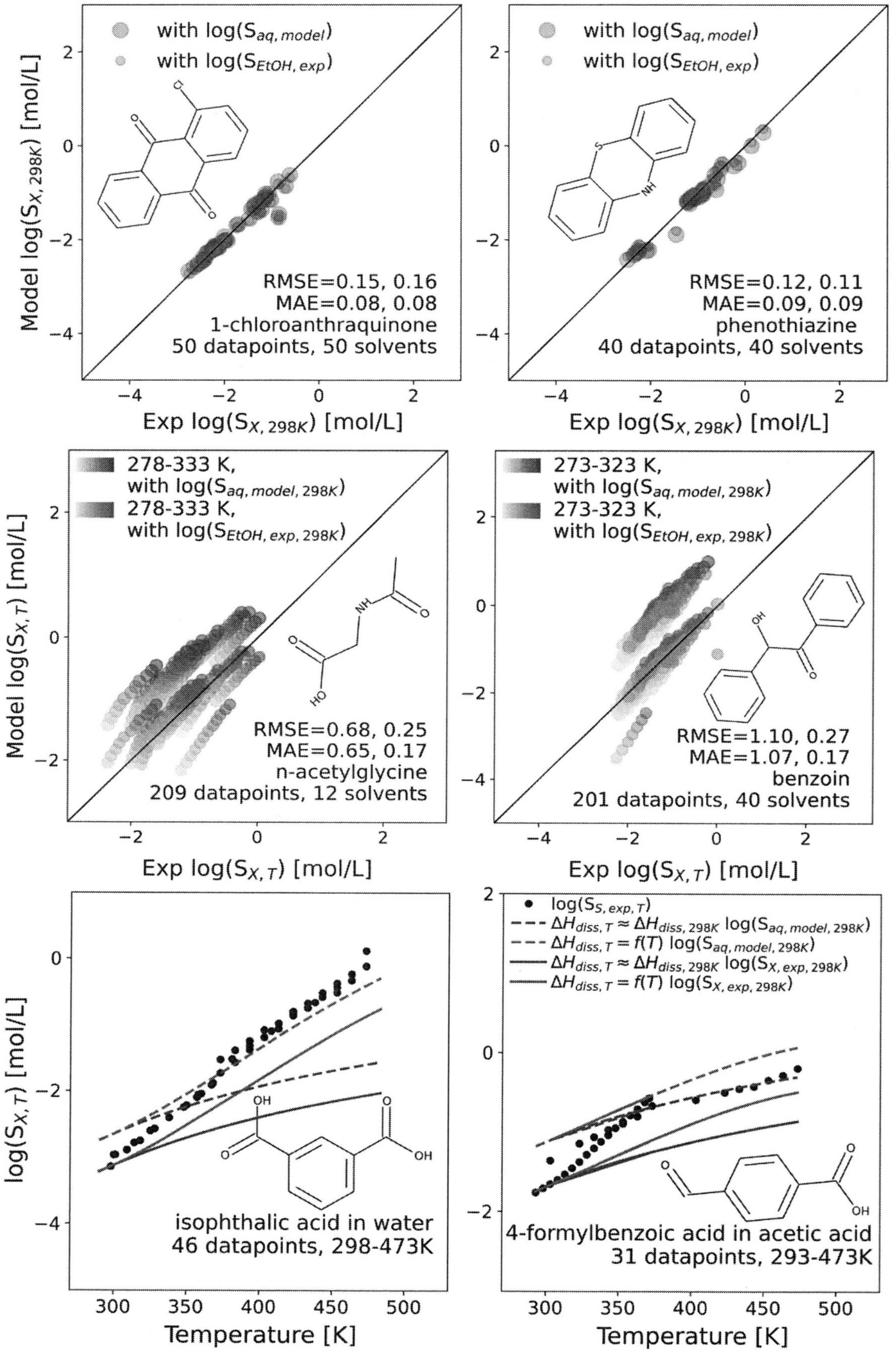
with log($S_{aq, model}$)
with log($S_{EtOH, exp}$)
Cl
O
O
RMSE=0.15, 0.16
MAE=0.08, 0.08
1-chloroanthraquinone
50 datapoints, 50 solvents
Model log($S_{X, 298K}$) [mol/L]
Exp log($S_{X, 298K}$) [mol/L]
with log($S_{aq, model}$)
with log($S_{EtOH, exp}$)
S
NH
RMSE=0.12, 0.11
MAE=0.09, 0.09
phenothiazine
40 datapoints, 40 solvents
Exp log($S_{X, 298K}$) [mol/L]
278-333 K,
with log($S_{aq, model, 298K}$)
278-333 K,
with log($S_{EtOH, exp, 298K}$)
NH
O
O
HO
RMSE=0.68, 0.25
MAE=0.65, 0.17
n-acetylglycine
209 datapoints, 12 solvents
Model log($S_{X,T}$) [mol/L]
Exp log($S_{X,T}$) [mol/L]
273-323 K,
with log($S_{aq, model, 298K}$)
273-323 K,
with log($S_{EtOH, exp, 298K}$)
OH
O
RMSE=1.10, 0.27
MAE=1.07, 0.17
benzoin
201 datapoints, 40 solvents
Exp log($S_{X,T}$) [mol/L]
OH
O
O
OH
isophthalic acid in water
46 datapoints, 298-473K
log($S_{X,T}$) [mol/L]
Temperature [K]
log($S_{S, exp, T}$)
$\Delta H_{diss,T} \approx \Delta H_{diss,298K}$ log($S_{aq, model, 298K}$)
$\Delta H_{diss,T} = f(T)$ log($S_{aq, model, 298K}$)
$\Delta H_{diss,T} \approx \Delta H_{diss,298K}$ log($S_{X, exp, 298K}$)
$\Delta H_{diss,T} = f(T)$ log($S_{X, exp, 298K}$)
O
O
OH
4-formylbenzoic acid in acetic acid
31 datapoints, 293-473K
Temperature [K]

297c Towards a Universal Adsorption Model Using Machine Learning: Motivation, Progress and Challenges

Diego Gomez Gualdron

Adsorption is central to innumerable processes of biological, physical and chemical nature. Hence, adsorption has concerned scientists for long. Perhaps more notably in the first half of the 20th century, when efforts to develop models to the predict the adsorption isotherm (the "heart" of adsorption studies) flourished. Some consider the Brunauer-Emmet-Teller (BET) model as the first attempt to create a universal adsorption theory, followed by models such as the Dubinin-Radushkievich (DR) model. However, it quickly became clear that an analytical equation including all factors affecting adsorption would have such complexity that it would be impractical, given the mathematical tools of the time.

More than 70 years later, adsorption continues to be critical in a diversity of applications. For example, for chemical separations, adsorption predictions are required to accelerate the discovery of adsorbents among a "search space" currently spanning millions of adsorbents, including "new age" materials such as metal-organic frameworks (MOFs). Although molecular simulation have helped navigate the above search space, the cost of adsorption predictions with this method has limited the "bandwidth" of the research community to a very limited number of chemical separations at specific conditions, while only considering limited points in the adsorption isotherm to estimate relative performance among adsorbents.

A fast universal model to predict adsorption isotherms would enable *i)* searching for suitable adsorbents for the ~80% of current chemical separations that could potentially see their energy cost reduced 10-fold by switching from a thermal method to an adsorptive method, and *ii)* coupling adsorbent and process optimization, as well as facilitating process modeling-based screening for the top adsorbents. Here it is proposed that now it is the time to retake the development of universal adsorption models by leveraging the mathematical tools of the "age of data." As a first step, it was demonstrated that the same deep neural network (DNN) trained with molecular simulation data for 200 alchemical molecules in 2,000 MOFs can predict individual full adsorption isotherms in the 0.01 to 100 bar range for CH_4, Ar, Xe, Kr, H_2, N_2, O_2, C_2H_6 in 400 MOFs. Moreover, the DNN model was found to have good transferability when tested on MOFs of a separate database previously unknown to the model.

To illustrate the screening possibilities that a universal DNN-based adsorption model opens, the model predictions were combined with ideal adsorption solution theory (IAST) to screen a ~50,000 MOF database for their capability to separate four industrially relevant mixtures (Xe/Kr, CH4/CH6, N2/CH4, and Ar/Kr) at multiple compositions and pressures. For this framework to work with sufficient accuracy, it was found critical for the DNN to make accurate predictions at low pressures (0.01–0.1 bar). After training a model with this capability, it was found that MOFs in the 95th and 90th percentiles of separation performance determined from DNN+IAST calculations were 65% and 87%, respectively, the same as MOFs in the simulation-predicted 95th percentile across several mixtures at diverse conditions (on average). Finally we discuss currently pursued strategies and related challenges in extending prediction capabilities of machine learning-based towards more diverse types of molecules.

297d Atomistic Simulations of RNA Folding in Bulk Aqueous Solutions and under Confinement

Gul Zerze

RNA is an essential molecule of life and is a promising agent to base drug and vaccine formulations. RNA is often encapsulated for pharmaceutical formulations, synthetic biology applications, and also for studying its evolution in the context of the origins of life. Encapsulation fundamentally alters thermodynamic conditions by reducing the available volume and modifying the chemistry of the surroundings. To investigate the effects of encapsulation on RNA folding, here we performed advanced sampling simulations of folding an explicitly solvated, all-atom GAGA tetraloop (10-nucleotides) encapsulated in model surfaces. We systematically modified the surface attractiveness and sizes of encapsulating sphere. We found that folded state is destabilized in the encapsulated RNA (with respect to the RNA in bulk aqueous solution) regardless of the attractiveness of the encapsulating surface or the size of encapsulation within the range we tested.

Effect of Free Volume Elements on the Local Dynamics of Glassy Polymers from Molecular Dynamics Simulations

Janani Sampath

Membrane separation is an economic alternative to energy intensive methods like distillation that is predominantly used in petrochemical industries for hydrocarbon separations. Polymers are attractive membrane materials owing to their inexpensive processing and scalability. An important feature of polymer membranes is the inefficient packing of bulky groups on the backbone that creates a distribution of connected void spaces known as free volume elements (FVEs). Glassy polymers - rigid and amorphous materials that have glass transition temperatures (T_g) well above room temperature, are widely used for gas and liquid separations. Recently developed glassy polymers like thermally rearranged polymers (TRPs) contain interconnected heterocyclic rings that further frustrate chain packing, leading to sub-nanometer voids. Despite the many advantages of glassy polymers, the phenomenon of plasticization – degradation of membranes from irreversible chain reorganization, has impeded the widespread implementation of these materials for industrial separation applications.

Motivated by the need to understand the molecular underpinnings of plasticization, we use molecular dynamics (MD) simulations to relate membrane structure to chain dynamics and flexibility. We perform all-atom MD simulations on three polymers with varying glass transition temperatures – polymethyl pentene, polystyrene, and thermally rearranged polymer (HAB-6FDA). We characterize FVEs using the alpha shape method and find that voids increase as a function of T_g and chain flexibility. We calculate the local dynamic fluctuations as a function of distance from FVEs using bond and dihedral autocorrelation functions; chain segments close to FVEs are more dynamic than chain segments away from the FVEs. We suggest means to reduce chain flexibility near FVEs, which will help mitigate plasticization while maintaining high separation efficiency.

297e Thermodynamic Stability of Multimetallic Nanoparticles

Giannis Mpourmpakis

Metal nanoparticles find tremendous application in modern technology, ranging from catalysis to medicine. However, how different metals mix at the nanoscale and their exact thermodynamically preferred chemical ordering (positioning of different metal atoms on the nanoparticle lattice) is unknown. In this presentation, I will highlight recent computational advances on accurately describing the thermodynamic stability of bimetallic nanoparticles of any size, shape, metal composition and chemical ordering, by combining Density Functional Theory calculations with machine learning. The developed models can be extended to multimetallic nanoparticles, as well as used for the description of chemical adsorption on the nanoparticle surface, thus, connecting nanoparticle stability with application properties (sensing, catalysis, etc.).

297f Probing Realistic Water-2D Material Interfaces Via Combined Quantum and Classical Simulations

Ananth Govind Rajan

Two-dimensional (2D) materials are gaining increasing attention for use in seawater desalination, osmotic power harvesting, and biological sensing devices. Because water comes into close contact with 2D material surfaces in these applications, it is important to understand water-2D material interactions at the molecular level. At the same time, 2D materials often have defects, such as vacancies and grain boundaries in them, and thus understanding their effect on interfacial properties is crucial to model such interfaces realistically. In this talk, I will highlight our recent work on the combined use of quantum-mechanical density functional theory (DFT) calculations, classical molecular dynamics (MD) simulations, and kinetic Monte Carlo (KMC) simulations to probe water-2D material interfaces. While DFT calculations can be used to predict the distribution of charge inside defective 2D materials, MD simulations can be used to simulate the thermodynamic and transport properties of interfacial water. Moreover, KMC simulations can be combined with DFT and graph-theoretic algorithms to predict the shapes of vacancy defects in 2D materials.

First, I will briefly outline how the combined application of KMC simulations, DFT calculations, and chemical graph theory has enabled the solution of the *isomer cataloging problem* for vacancy defects in 2D materials. I will demonstrate the excellent agreement of simulated vacancy defect shapes in two prototypical 2D materials – graphene and hexagonal boron nitride (hBN) – with experimental transmission electron microscopy images. Not only that, I will also briefly discuss the use of machine learning to develop structure-formation relationships for vacancy defects in graphene.

Second, I will discuss the effect of vacancy defects and surface roughness on the wettability and slip length of water on hBN. Indeed, at the molecular level, the no-slip boundary condition is violated, as quantified by the slip length of water on the surface. I will show that vacancies at a lower concentration of 0.082 nm^{-2} do not affect the wettability of hBN, although they still affect the water slip length. On the other hand, vacancies at a larger concentration of 0.32 nm^{-2} affect interfacial properties significantly. In fact, nitrogen vacancies at such concentrations can increase the slip length of water on hBN threefold to around 18 nm, presenting hBN as an alternative high-slip surface to graphene. I will also explain how surface roughness in hBN can explain the water contact angle of 66° and water slip length of 1 nm measured experimentally, simultaneously highlighting the prominent role played by electrostatic interactions in the interfacial properties of water on realistic hBN surfaces.

Third, I will present a study on the role of grain boundaries and interfacial electrostatic interactions in modulating water and ion transport via bicrystalline nanoporous hBN. I will discuss how grain boundaries alter the areas and shapes of nanopores in bicrystalline hBN, as compared to the nanopores in monocrystalline hBN. Although bicrystalline nanoporous hBN with a lower misorientation angle shows an improved water flow rate by ~30%, it demonstrates reduced Na^+ ion rejection by ~6%, as compared to monocrystalline hBN. I will also explain the role of the nanopore shape in water desalination, demonstrating that more elongated pores with smaller sizes can match water permeation through less elongated pores of slightly larger sizes, with a concomitant ~3–4% decrease in Na^+ rejection. Finally, I will discuss how the water flow rate is affected by interfacial electrostatic interactions. In this regard, the water flow rate is the highest when altered partial charges on B and N atoms are determined using DFT calculations, as compared to when no partial charges or bulk partial charges (i.e., charged hBN) are considered.

Overall, these multi-scale investigations of thermodynamic and transport properties offer new insights into the wettability of defective 2D material surfaces and water flow on them, as well as into water and ion transport through nanopores in realistic 2D materials.

711A Correlating and Predicting Thermodynamic Properties of Fluid Mixtures in Phase Equilibrium and Adsorption Equilibria

Chau-Chyun Chen

Accurate correlation and prediction of thermodynamic properties of fluid mixtures is essential for industrial process simulation, design, and optimization. The classical nonrandom two-liquid (NRTL) activity coefficient model for nonelectrolyte solutions[1] and the electrolyte NRTL model for electrolyte solutions[2] have been considered among the most versatile and widely practiced molecular thermodynamic models to correlate and predict liquid phase nonideality of multicomponent liquid mixtures. However, without explicitly accounting for specific chemical associations such as hydrogen bonding and ion hydration, these NRTL models fail to accurately correlate and predict phase behavior of highly associating nonelectrolyte systems such as methanol – alkanes binaries and aqueous electrolyte solutions with highly charged ionic species such as lithium and magnesium salts. This paper present how the recently developed association NRTL model and association electrolyte NRTL model address these issues for many highly challenging and impactful multicomponent association systems.[3,4,5] This paper further reports recent successful extension of the NRTL model for the adsorbed phase nonideality in multicomponent mixed-gas and liquid adsorption equilibria.[6,7]

References

1. H. Renon, J.M. Prausnitz, Local Compositions in Thermodynamic Excess Functions for Liquid Mixtures, *AIChE Journal*, **1968**, 14, 135-144.
2. Y. Song, C.-C. Chen, Symmetric Electrolyte Nonrandom Two-Liquid Activity Coefficient Model, *Industrial andamp; Engineering Chemistry Research*, **2009**, *48*, 7788-7797
3. Y. Hao, C.-C. Chen, Nonrandom Two-Liquid Segment Activity Coefficient Model with Association Theory, *Industrial andamp; Engineering Chemistry Research*, **2019**, 58, 12773-12786.
4. Y. Hao, C.-C. Chen, Nonrandom Two-Liquid Activity Coefficient Model with Association Theory, *AIChE Journal*, **2021**, 67, e17061.
5. Y. Lin, C.-J. Hsieh, C.-C. Chen, Association-based Activity Coefficient Model for Electrolyte Solutions, *AIChE Journal*, **2022**, 68(2), e17422.
6. H. Kaur, H. Tun, M. Sees, **C.-C. Chen**, "Local Composition Activity Coefficient Model for Mixed-Gas Adsorption Equilibria," *Adsorption*, **2019**, 25 (5), 951-964.
7. U. Hamid, P. Vyawahare, H. Tun, **C.-C. Chen**, "Generalization of Thermodynamic Langmuir Isotherm for Mixed-gas Adsorption Equilibria," *AIChE Journal,* **2022,** 68(6), e17663

711B Machine Learning for Predicting the Viscosity of Binary Liquid Mixtures

Camille Bilodeau, Andrei Kazakov, Sukrit Mukhopadhyay, Jillian Emerson, Tom Kalantar, Chris Muzny, Regina Barzilay, Klavs Jensen

Viscosity is a critical parameter in process engineering and a key design objective for application areas ranging from coatings and lubricants to the personal care and pharmaceutical industries. The lack of reliable and general methods for predicting the viscosities of mixtures creates a barrier for formulation design and process engineering. In this work, we developed a graph-based neural network architecture and applied it to the problem of predicting the viscosity of binary liquid mixtures as a function of composition and temperature. To obtain a high-quality training dataset, we developed an automated curation pipeline and applied it to a large dataset collected from the literature by the National Institute of Standards and Technology (NIST) to be used as training data. The resulting model accurately predicts viscosity, with an MAE of 0.043 and an RMSE of 0.080 in log cP units. To improve the dependability of the model in application settings, we also developed a "reliability evaluator" that determines whether or not a prediction is reliable based on the variance between an ensemble of models. For predictions considered "reliable" (80% of the test set), the model performs sufficiently well that its error falls within the experimental uncertainty of the training data, with an MAE of 0.029 and an RMSE of 0.047. Overall, this work provides 1) a large set of curated viscosity data that can be used for future machine learning efforts, 2) a new, graph-based deep learning approach for predicting the viscosity of binary mixtures that inf, and 3) an illustrative case study for understanding the role that deep learning can play in achieving accurate and reliable property prediction.

711C Computational evaluation of liquid mixtures using physics-based models and machine learning

Mohammad Atif Faiz Afzal, Garvit Agarwal, Andrea Browning, Mathew D. Halls

Complex fluid mixtures are critical components in many industrial processes and formulations. Understanding the physical properties of such fluid mixtures is vital in optimizing the process parameters, conditions, and compositions. An interesting feature of such complex mixtures is that there are numerous combinations, however it is impractical to test a large number of candidates in lab. Predictive models are therefore beneficial to obtain insights into the behavior of liquid mixtures. Traditional mixing rules often fail to capture the mixture properties as they do not account for inter-molecular interactions. One approach to overcome this shortcoming is to use molecular dynamics (MD) simulations. Typical MD simulations consider constant partial charges of atoms, and furthermore, do not consider polarization leading to inaccurate property estimations. To capture the correct partial changes, polarization, and dynamics, we use machine-learned (ML) potentials trained using thousands of ab-initio data. The resulting MD using such ML potential is highly accurate and can be extended to highly complex fluids, including ionic liquids. In this presentation, I will demonstrate the application of ML potential for liquid mixtures of electrolytes and ions and compare it with experimental results and classical MD.

711D Predicting Properties of Ionic Liquid Mixtures Using Molecular Simulations and Machine Learning

Jindal Shah

Ionic liquids are composed exclusively of an organic cation and anion that can be either organic or inorganic in nature. Highly articulated nature of the ions implies a delocalization of either the positive or negative charge, resulting in melting points of these compounds below ambient conditions. Extremely low vapor pressures and ability to target properties suitable for a given application by carefully selecting the ion combination are the primary drivers for research on the application of ionic liquids in CO_2 capture, gas separations, batteries, solar cells, and reactions, etc. While designing an ionic liquid, one is confronted with the vast chemical space for selecting cations and anions; it is estimated that there are more than a billion ionic liquids. Considering that almost all applications of ionic liquids are likely to involve mixtures of ionic liquids or ionic liquid-organic solvents, the number of such possible combinations would be in trillions. High throughput screening using machine learning and understanding molecular phenomena with atomistics simulations play an important role to navigate the vastness of the available chemical space. In this presentation, challenges and opportunities for predicting properties of ionic liquid mixtures from atomistic simulations will be highlighted. We will also discuss our efforts at employing machine learning in rapidly estimating properties of ionic liquid-ionic liquid mixtures and gaps that must be bridged for property predictions.

711E A Multi-scale Computational Framework for Property Prediction of Fluid Mixtures

Marianna Yiannourakou, Xavier Rozanska, Benoit Minisini

Fluids are present in many industrial products and processes, with applications ranging from upstream oil and gas exploration and production to fuels and additives as well as pharmaceuticals and drug delivery. Improvements in industrial processes currently rely on accurate modeling of heat and fluid flow modeling requiring reliable data as numerical input. Due to the difficulties to get sufficient accuracy on the experimental data over certain ranges of temperatures and pressure, accurate estimates of physical and chemical properties of fluids increasingly make use of powerful atomistic simulation and multi-scale techniques.

The term multi-scale modeling comprises varying methodologies and levels of theory[1], starting at the molecular level and reaching "up" to continuum methods and equations of state.

The physical and chemical characteristics of a material, which relate to a given property or process, determine the time and length scales at which simulations need to occur. In turn, the property- or process-relevant length and time scales require simulations using a certain level of theory, such as quantum or statistical mechanics. Hence, for all practical applications, simulations proceed by decomposing a complex system or process into parts or steps, which are studied individually. Thus, gained knowledge and data are then combined to predict the behavior of the full system or process.

Here, we will illustrate (through the use of efficient simulation workflows) the value of simulations at different scales for determining a broad range of properties including vapor-liquid equilibria (VLE)[2], solubility[3], surface tension, viscosity, thermal conductivity, and heat capacity, for organic fluids of industrial relevance. We will illustrate efficient protocols for deriving simple and easy-to-use correlations for properties of interest[4], based on large datasets obtained by simulation.

Specific examples will be used to illustrate the accuracy, application range, and limit of the underlying computational approaches and provide perspectives and development trends.

1. Nieto-Draghi, C.; Fayet, G.; Creton, B.; Rozanska, X.; Rotureau, P.; de Hemptinne, J.-C.; Ungerer, P.; Rousseau, B.; Adamo, C., A General Guidebook for the Theoretical Prediction of Physicochemical Properties of Chemicals for Regulatory Purposes. *Chemical Reviews* **2015,** *115* (24), 13093-13164.
2. Yiannourakou, M.; Ungerer, P.; Lachet, V.; Rousseau, B.; Teuler, J. M., United atom forcefield for vapor-liquid equilibrium (VLE) properties of cyclic and polycyclic compounds from Monte Carlo simulations. *Fluid Phase Equilib.* **2019,** *481*, 28-43.
3. Yiannourakou, M.; Rozanska, X.; Minisini, B.; de Meyer, F., Molecular simulations for improved process modeling of an acid gas removal unit. *Fluid Phase Equilib.* **2022,** *560*, 113478.
4. Ungerer, P.; Yiannourakou, M.; Mavromaras, A.; Collell, J., Compositional Modeling of Crude Oils Using C10–C36 Properties Generated by Molecular Simulation. *Energy andamp; Fuels* **2019**.

46a Characteristics of Droplet Explosions Studied with Non-Equilibrium Molecular Dynamics Simulations

Dominik Schaefer, Babette Kunstmann, Maximilian Kohns, Hans Hasse

In flame spray pyrolysis, functional nanoparticles are produced by dissolving a metal-containing precursor in a combustible solvent and spraying the mixture into a flame. In that process, the droplets of the precursor solution undergo a cascade of microexplosion events that originate from within the droplet or close to its surface [1-3]. Interestingly, the quality of the obtained nanoparticles is strongly related to the occurrence of such microexplosions [2,3]. However, to date, the causes for these microexplosions are not well understood, mainly because experimental investigations of the interior of the droplets are infeasible. It is suspected that the microexplosions are linked to internal superheating in the droplet [1-3].

In this work, we have developed a non-equilibrium molecular dynamics simulation scenario to investigate characteristics of droplet explosions on a fundamental level. The simulation scenario was set up and tested considering the *Lennard-Jones truncated and shifted* (LJTS) model fluid, since it is a realistic model of the qualitative behavior of fluids and computationally inexpensive. In the simulation scenario, a droplet of a pure fluid is first equilibrated with its coexisting vapor. Then, in the middle of the droplet, a thermostat is activated that imposes a high temperature. Several characteristics of the temporal evolution of the system are then tracked, e.g. the position of the droplet-gas-interfaces, the number of particles in the liquid phase, and the radius of gyration of the droplet. Using that simulation scenario, a series of simulations was carried out in which the set temperature of the thermostat was varied. For low set temperatures, the droplet continuously evaporates, following the d^2 law of droplet evaporation [4]. By contrast, for high set temperatures, a vapor bubble nucleates inside the liquid droplet, the droplet expands and eventually breaks up. This process also involves several oscillations of the droplet size. These qualitative results help to identify causes for droplet microexplosions.

[1] N. Jüngst et al., Exp. Fluids 63 (2022) 60.
[2] C. Rosebrock et al., AIChE J. 62 (2016) 381-391.
[3] H. Li et al., Combust. Flame 215 (2020) 389-400.
[4] F. Meierhofer and U. Fritsching, Energ. Fuel. 35 (2021) 5495-5537.

46b Bubble Nucleation in the Surfactant Stabilized Polyol-CO_2 Mixtures:Insights from a Classical Density Function Theory Study

Sriteja Mantha, Huikuan Chao, Andrew Ylitalo, Benjamin Laccetti, Thomas Fitzgibbons, Weijun Zhou, Valeriy Ginzburg, Richard C. Flagan, Julia A. Kornfield, Zhen-Gang Wang

Polyol based foams are widely sought-after materials for manufacturing thermal insulators, high resilience foam seating, adhesives, hard plastics for electronic instruments, etc. The polyol foams are produced by the reaction of di-isocyanate with polyol to form polyurethane and water. Some amount of isocyanate reacts with water to produce CO_2. The generated CO_2 nucleates into bubbles within the polymer matrix, forming a foam. Silicone-polyether(SPE) surfactants, made of poly dimethyl-siloxane backbone and polyether branches, are commonly used to stabilize the foam formulation. The SPE surfactants reduce interfacial tension between polyol-CO_2 interface, promotes bubble generation and impacts the foam cell size. It is also known that the composition of SPE surfactant significantly influences its role in stabilizing the foam formulation as well. However, the physical mechanism of how these SPE surfactants affect the nucleation and stability of the bubbles is not well understood. In this talk, using classical density functional theory models, we propose design principles for SPE surfactants and elucidate the mechanism through which they lead to foams with improved physical properties.

46c Interfacial Thermodynamics of Cryogenic Fluids: The Effect of Non-Condensable Gas on Fluid Storage

Michael DeLyser, Ashwin Ravichandran, Wayne J. Mullinax, John W. Lawson

Propellant tanks that contain cryogenic fluids (CFs) in low gravity conditions are usually pressurized with non-condensable (NC) gases for fast extraction. Unfortunately, the presence of NC gases causes CFs to exhibit higher boil-offs compared to pure CF systems. For optimal utilization of the cryogenic fuels, these higher boil-off rates must be minimized. Our goal is to quantify the effects NC gases have on the evaporation and condensation dynamics of CFs to develop strategies that mitigate the higher boil-offs. The hypothesis is that the NC gas accumulates around the CF's liquid vapor interface in the Knudsen layer, creating a kinetic barrier for both evaporation and condensation. Testing this hypothesis with experimental methods is difficult due to the transient nature of the Knudsen layer, which is only a few nanometers thick. Furthermore, experimental capabilities are limited in low gravity conditions and extremely expensive. Accordingly, we approach this problem from a theoretical perspective. In this study, we employ molecular dynamics (MD) simulations to probe the interfacial mechanisms that affect CF's evaporation and condensation at varying concentrations of NC gases. Specifically, we use nitrogen (N_2) and oxygen (O_2) as our CFs and neon (Ne) as our NC gas. Using MD simulations, we show that Ne accumulates at N_2 liquid vapor interface across a wide range of Ne concentrations, thereby impeding mass transport of N_2. Our simulations allow for direct computation of the molar flux as well as the mass accommodation coefficient (MAC) which can then be used as input parameters to continuum fluid dynamics (CFD) models for optimal storage tank design.

46d Energy Optimization of Nanochannel Fluid Extraction with Various Channel Geometries

Zachary Diermyer, Yidong Xia, Jiaoyan Li

Many factors influence the extraction of hydrocarbons from shale and other subsurface sources. During secondary oil recovery, an injection well floods the underground reservoir with inexpensive fluids like seawater. Occasionally, a chemically engineered working fluid is created using surfactants, and injected instead. Attempts have been made to computationally determine the role salt content and surfactants play in this process. None of these studies directly targeted the relative wettability between water and hydrocarbons, while simultaneously accounting for their nanoconfinement within solid channels of varying surface morphologies. DPD is a coarse-grained, particle-based computational model capable of performing mesoscale simulations. Thus, DPD can give us insights into hydrocarbon recovery at tens of nanometer length scales and for tens of nanosecond time scales. This cannot be accomplished with molecular dynamics models using current computational resource capabilities. We perform many-body dissipative particle dynamics (mDPD) simulations to study fluid flow in channels with sizes ranging from 4.5 nm to 13.5 nm. Model parameters are rigorously calibrated to handle the realistic working fluid, water, and source fluid, heptane (C_7H_{16}), confined in amorphous silica ($aSiO_2$) channel walls. Different pressure levels are applied to the water as pressure driven boundary conditions, and heptane is extracted from the channel. The mDPD attractive parameter is varied in order to change water from a wetting fluid to a relatively non-wetting fluid. Based on the study of channel geometries (i.e., flat, convergent-divergent, and pocket designs), we find a relationship between increased surface roughness and permeability reduction. We will present an average power exerted on the water during initial channel flushing, as well as the maximum amount of heptane recovered, as a function of the water wettability and surface roughness. An optimization algorithm will be proposed to determine the most efficient heptane extraction forcing term in each case.

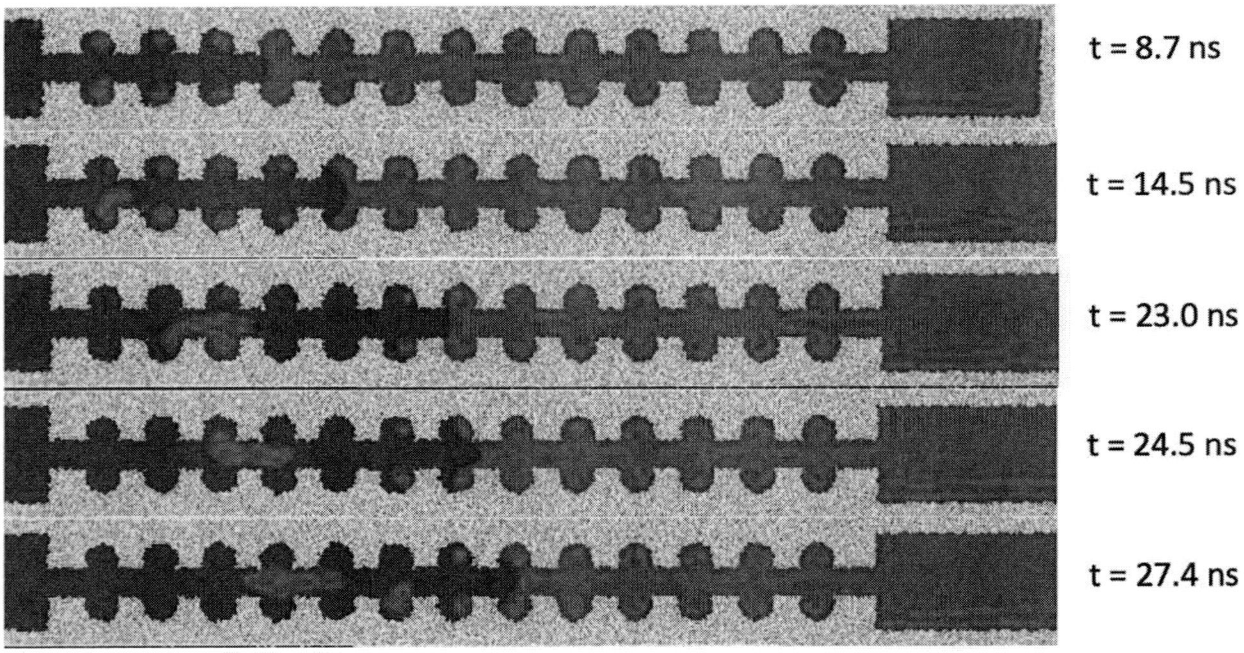

Channel Length: $L_C = 125$ nm

46f Free-Energy of Monomeric Species in Uio-66 Metal-Organic Framework

Sanoj -, Yamil Colón

Metal-organic frameworks (MOFs) have been explored in gas separation, purification and increasing hybrid-MOFs polymer systems due to their pore surfaces environment and distribution[1]. Pore surface's environment and their interaction with different molecules can help understand the physical properties of hybrid MOFs-polymer systems. In our work, we have focused on understanding the interaction between the pores of UiO-66 as shown in Fig. 1A, and different monomers, as shown in Fig. 1B, which are similar in structure but have different functional groups. We have used advanced sampling techniques to calculate the free energy profile of monomers traversing through the pores. Figures 1C and 1D show the different pathways of monomers traversing through the pores and the free energy profiles. We have observed that the free energy profile is very different for different monomers and is dependent on the interaction of these monomers with the pore environment. The free energy profile can help explain the physical properties of the hybrid MOFs-polymer system and the stability of these hybrid-MOFs polymer systems.

Reference:

1. Liang, B et al, J. Am. Chem, Soc. 2022, 142, 17795-17801.

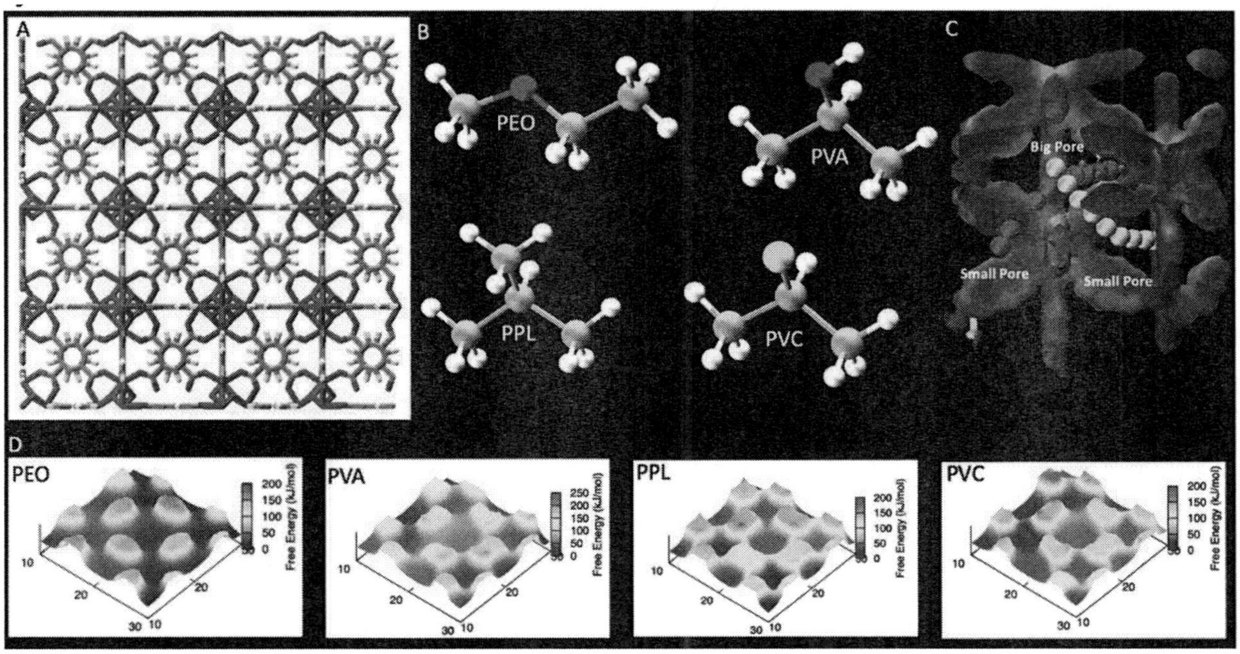

Figure 1: (A) UiO-66 MOFs. (B) Different monomers are Polyethylene oxide (PEO), Polyvinyl alcohol (PVA), Polypropylene (PPL), and Polyvinyl chloride (PVC). (C) Finite temperature string (FTS) of monomers going from one pore to another. (D) Free energy profile of different monomers at 300 K.

46g Structural Origin of Multiple Alkylated Cyclopentane As an Effective Lubricant

Jee-Ching Wang, Vidit Singh

Multiple Alkylated Cyclopentanes (MACs) have been utilized as a new class of effective lubricant recently for various applications including NASA's space missions. In this study, the molecular dynamics (MD) method coupled with appropriate potential models has been used to simulate MAC molecules and construct realistic thin films sandwiched between two a-alumina surfaces. The results from the MD simulations show that, unlike symmetric molecules that can be induced by nanoscopic confinement to form layered configurations, the highly branched MAC molecules possess sufficient structural asymmetry and complexity to resist the layering tendency. Molecular interactions with the surfaces are found to turn the branches into non-trans conformations to allow parallel orientation with respect to the surfaces so that the interactions are enhanced and able to function as anchor to stabilize non-layered configuraions of the branches in a crowed environment. For comparison, the branches are severed from the cyclopentane core and the resultant mixture film becomes much more layered. These structural properties underlie the effectiveness of MACs as a new class of lubricant and provide a scientific foundation for future lubricant development.

46h Extracting Anisotropy Strength and Interfacial Free Energy of Al-Mg Alloy Under Rapidcooling Conditions Using Molecular Dynamics Simulations

Daniel Dolce, Pabitra Choudhury

Many pure metals and alloys produced in industry suffer from manufacturing defects, such as the formation of microstructural branching defects called dendrites. The microstructure is strongly affected by solid liquid interfacial properties during solidification process. The anisotropy of the interfacial free energy (γ) has an especially strong impact on both the tip radius and growth direction of dendrite defects. It has been previously shown that the composition and undercooling temperature of both pure metal and binary alloys can strongly affect the anisotropy and cause the dendrites to change growth directions. Due to their wide use in industry, understanding these effects in aluminum-based alloys are of great interest. Specifically, the Al-Mg binary alloy was studied at various undercooling temperatures to determine if a possible temperature dependence of the anisotropy of γ exists. Molecular Dynamics (MD) simulations were performed in the Large-Scale-Atomic/Molecular Massively Parallel Simulator (LAMMPS) using a Modified Embedded Atom Method (MEAM) interatomic potential from the NIST IPR database. The solid-liquid interface was identified using an order parameter, which classifies atoms with nearest neighbors in both the solid and liquid states. The Fast Fourier Transform (FFT) and wavenumber (k) were used to quantify the rough behavior of this interface and extract the interfacial stiffness. This was done for the (100) and (110) orientations at undercooling temperatures of 910, 850, and 750K and the anisotropic parameters ε_1 and ε_2 were determined. Our results show that the interfacial free energy typically followed a relationship of $\gamma_{100} > \gamma_{110} > \gamma_{111}$. Except for 910K which followed a $\gamma_{110} > \gamma_{100} > \gamma_{111}$ relationship. Our results further predict that the dendrite growth in the (100) direction at 910K followed by a switch to a hyperbranched direction at 850K, followed by a return to the (100) direction at 750K.

Acknowledgements: The work is supported by ARL Grant No. W911NF-2020032 and used the Extreme Science and Engineering Discovery Environment (XSEDE) TACC at the stampede2 through allocation [TGDMR140131]. This work utilized resources from the University of Colorado Boulder Research Computing Group, which is supported by the National Science Foundation (awards ACI-1532235 and ACI-1532236), the University of Colorado Boulder, and Colorado.

46i Dynamic Evolution of Atomically Dispersed Catalysts

Nicholas Humphrey, Selin Bac, Shaama Mallikarjun Sharada

Atomically dispersed catalysts have become popular due to the benefits gained by ease of separation due to the two phases of heterogeneous catalysts while maintaining the selectivity and efficiency of homogenous catalysts. However, the coordination environment and diffusivity of precious metal atoms on support surfaces are still not fully understood by experimental and static Density Functional Theory (DFT) studies. Sintering and dynamics of the support surface have been reported to be affected by conditions such as reactant partial pressures and reaction temperature, leading to the importance of studying the dynamical evolution of these systems. We utilize picoseconds timescale Ab Initio Molecular Dynamics (AIMD) simulations to better understand the coordination environment, diffusivity, and stability of Pt over rutile TiO_2 (110) support interface. With elevated temperature (1200K) 2ps AIMD simulations, we uncover eight new binding sites of Pt on the TiO_2 surface. To the best of knowledge, these sights have not been reported in static DFT studies because these structures are not intuitive and AIMD allows us to sample many configurations. These configurations have been selected based on criteria such as coordination number of Pt to the surface, migration events, and Bader charge of the Pt and nearby surface atoms. We have studied the dynamics of eight common Water-Gas Shift (WGS) adsorbates bound to both the Pt and support surface with lower temperature (500K) 5ps AIMD simulations. During these simulations, we observe the Pt diffuse to the unstable bridge sites of the surface while bound to COOH and CO_2, Pt diffusion along the basal plane via hydrogen bonding of OH to the support O atoms, and stabilization due to the formation of linear O-Pt-X complexes where O belongs to the support surface and X is the adsorbate during the simulation. We plan to analyze the chemical bounding of these near linear complexes to better understand what causes Pt stabilization, Pt oxidation state and how that determines catalytic activity, and uncover adsorbates that may induce/inhibit metal atom migration and target specific oxidataion states.

46j Molecular Simulations Study of Calcium Carbonate-Amino Acid-Dentin Interactions

Alina Emelianova, Max Maximov, Tatiana V. Brinzari, Andrei Potanin, Gennady Gor

Dentin hypersensitivity is a common dental health condition usually associated with exposed dentin surfaces and defined as a short, sharp pain in response to chemical or thermal stimuli. One of the resolutions for this problem is addressed by Colgate® Sensitive Pro-ReliefTM dentifrice by providing physical occlusion of exposed dentinal tubules by aggregation and adhesion of calcium carbonate abrasive particles. Amino-acids presence in toothpaste, such as arginine and lysine, are expected to help the adhesion between the dentine surface and the abrasive particles and thus provide better dentinal tubules occlusion. However, the specific mechanism of interaction of amino-acid molecules with the abrasive particles and dentin components is not clear. Moreover, quantifying and direct comparison of adhesion effect on the surfaces of dentine and calcium carbonate provided by the inclusion of different amino acids has not been done.

In this work, we investigate the interactions of amino acids and calcium carbonate (calcite), and dentin (represented by hydroxyapatite, HAP) surfaces through classical molecular dynamics simulations. To quantify the adhesion effect, we estimate the potential of mean force (free energy profiles) of the separation of the calcite and HAP plates in pure water and in a mixture of water and amino acid. Simulations showed that the free energy change is noticeably higher in the case of the presence of amino acids in the system, and differs depending on the type of amino acid. The results indicate that the higher adhesion between calcite and HAP surfaces can be achieved by the inclusion of arginine components into the toothpaste solution, thus easing the blocking of exposed dentin tubules and providing effective protection against the external irritants.

656a A New Computational-Experimental Screening Methodology Identifies More Effective Solvents for CO_2 Capture

Frederick de Meyer, Alexey Orlov, Christophe Coquelet, Xavier Rozanska

Carbon capture and storage technologies are projected to increasingly contribute to cleaner energy transitions by significantly reducing CO_2 emissions from fossil fuel-driven power and industrial plants. The state-of-the-art technology for CO_2 capture is the chemical absorption with aqueous alkanolamines. The important equipment cost and parasitic energy due to the thermal regeneration of the solvent prevents its widespread application. Unfortunately, solvents showing a lower regeneration energy tend to absorb CO_2 more slowly, resulting in much larger equipment. Therefore, new industrial solvents are often a mixture of amines with relatively low regeneration energy (e.g. tertiary amines) and an activator, piperazine, which increases the overall CO_2 absorption rate. The search for new tertiary amines with better kinetics is thus of paramount importance. Improving the efficiency of experimental screening using computational tools is challenging due to the complex nature of chemical absorption. We have developed a novel computational approach that combines kinetic experiments, molecular simulations and machine learning for the *in silico* screening of hundreds of prospective candidates and identify a class of tertiary amines that absorbs CO_2 faster than a typical commercial solvent when mixed with piperazine, which was confirmed experimentally.

Figure 1 illustrates the workflow of the methodology. First, we have developed a computational approach based on molecular modeling to predict the CO_2 absorption rates and energies in aqueous tertiary amines. The predictions are validated using experimental data. The key to the new model's excellent accuracy is to include the solvation effects in the computation of the free energy barrier for the reaction of CO_2 absorption [1].

In a second step we have generated a diverse library of 100 tertiary amines which represent a large portion of chemotypes relevant for industrial amine-based chemical CO_2 absorption. Using the molecular simulations-based model, we have calculated the CO_2 absorption rates and energies for those 100 aqueous amine solvents. As a check, some points have been verified experimentally [2].

In a third step we have used the dataset of the CO_2 absorption rates and energies for the 100 aqueous amines to build quantitative structure-property relationship models. These models allowed us to perform a large-scale virtual screening of commercially available amines and to identify the amines with optimal CO_2 absorption rate and energy. For several amines the predictions have been verified experimentally. Subsequently, for the best performing amine, piperazine has been added. The approach and the results will be presented.

[1] Rozanska, X., Wimmer, E. andamp; de Meyer, F. Quantitative Kinetic Model of CO_2 Absorption in Aqueous Tertiary Amine Solvents. *J. Chem. Inf. Model.* 61, 1814–1824 (2021)

[2] Orlov, A., Valtz, A., Coquelet, C., Rozanska, X., Wimmer, E., de Meyer, F. Computational screening methodology identifies effective solvents for CO_2 capture. *Communications Chemistry* 5, 37 (2022)

656b Characterization of Chemoresponsive Liquid Crystals Using Topological Descriptors and Machine Learning

Shengli Jiang, Nanqi Bao, Alexander Smith, Nicholas L. Abbott, Victor Zavala

Chemoresponsive liquid crystals (LCs) provide a versatile materials platform for designing affordable and lightweight gas (analyte) sensors [1]. Specifically, LCs can be designed to change their orientation ordering and optical birefringence when exposed to a certain target chemical environment [2]. The optical response of an LC system contains a rich set of features (e.g., space-time color and brightness patterns) that result from multiple physical processes taking place in the LC film, on the supporting surface, and at different scales (e.g., diffusion of the analyte through the film and binding/reaction events occurring at the surface) [3].

Machine learning techniques have been recently used for characterizing the optical response of LC crystals and to infer analyte species and concentrations from such responses. For example, Smith et. al. analyzed the spatial response of LCs when exposed to dimethyl methylphosphonate (DMMP) or water. The authors extracted response features using filters obtained from a pre-trained convolutional neural network (CNN) and showed that such features can yield a near-perfect classification accuracy of the chemical environment [4]. However, while NN-based feature extraction is powerful, this approach also has some important drawbacks. For example, due to the "black box" nature of CNNs, there is limited explainability for the response features that they are searching for. In addition, since CNNs contain large numbers of parameters, they are prone to overfitting and can be difficult to train.

Recent work has also used topological descriptors to characterize LC responses. Topological descriptors, in particular the so-called Euler Characteristic (EC) and the fractal dimension (FD), have been traditionally used to characterize topological defects in LCs [5]–[9]. For example, Muniandy et al. studied the birefringence textures of isotropic lamellar LC systems using fractal dimension [10] and established a correlation between the FD of the birefringence texture and the morphological structure of the LC system. Smith et. al. applied EC to characterize simulated micrographs for nematic LC systems [11] and demonstrated that EC outperformed traditional tools such as the Fourier transform and Moran's I in distinguishing between LC systems. Solis et. al. used persistent homology to track the structural changes in a LC nanocomposite and revealed the effect of confined geometry on the nematic-isotropic and isotropic-nematic phase transition [12].

In this study, we provide a unifying view of various topological descriptors through the lens of Minkowski functionals (MFs), which describe diverse geometric and topological properties such as shape, convexity, and connectivity [13]. MFs have been traditionally applied to binary fields (e.g., two-phase materials) and thus provide single scalar descriptors. Unfortunately, this approach requires manually selecting a threshold value to binarize the field. Here, propose to use filtration techniques to select multiple threshold values across the domain and compute a series of MF values [14]. Filtration enables capturing of richer topological information from the fields. In addition, we discuss connections between topological descriptors and Gaussian random fields, which enables interpretability of spatial features from a mechanistic perspective. We demonstrate the developments using a couple of case studies; here, we analyze the spatial responses of LC to sulfur dioxide and water, as well as the space-time responses to ozone and chlorine. We compare topological descriptors to state-of-the-art CNNs and show that such descriptors can achieve comparable prediction accuracy but at a much lower computational cost and with higher degree of interpretability.

References:

[1] R. R. Shah and N. L. Abbott, "Principles for Measurement of Chemical Exposure Based on Recognition-Driven Anchoring Transitions in Liquid Crystals," Science, vol. 293, no. 5533, pp. 1296–1299, 2001, doi: 10.1126/science.1062293.

[2] R. J. Carlton et al., "Chemical and biological sensing using liquid crystals," Liquid Crystals Reviews, vol. 1, pp. 29–51, 2013, doi: 10.1080/21680396.2013.769310.

[3] J. T. Hunter and N. L. Abbott, "Dynamics of the chemo-optical response of supported films of nematic liquid crystals," Sensors and Actuators, B: Chemical, vol. 183, pp. 71–80, 2013, doi: 10.1016/j.snb.2013.03.094.

[4] A. D. Smith, N. Abbott, and V. M. Zavala, "Convolutional Network Analysis of Optical Micrographs for Liquid Crystal Sensors," Journal of Physical Chemistry C, vol. 124, no. 28, 2020, doi: 10.1021/acs.jpcc.0c01942.

[5] A. D. Kiselev, R. G. Vovk, R. I. Egorov, and V. G. Chigrinov, "Polarization-resolved angular patterns of nematic liquid crystal cells: Topological events driven by incident light polarization," Physical Review A - Atomic, Molecular, and Optical Physics, vol. 78, no. 3, 2008, doi: 10.1103/PhysRevA.78.033815.

[6] E. Pairam et al., "Stable nematic droplets with handles," Proceedings of the National Academy of Sciences of the United States of America, vol. 110, no. 23, 2013, doi: 10.1073/pnas.1221380110.

[7] J. W. Tsung, Y. Z. Wang, S. K. Yao, and S. Y. Chao, "Crystal-like topological defect arrays in nematic liquid crystal," Applied Physics Letters, vol. 119, no. 12, 2021, doi: 10.1063/5.0064303.

[8] A. Fernández-Nieves et al., "Novel defect structures in nematic liquid crystal shells," Physical Review Letters, vol. 99, no. 15, 2007, doi: 10.1103/PhysRevLett.99.157801.

[9] I. Muševič, "Nematic liquid-crystal colloids," Materials, vol. 11, no. 1. 2017. doi: 10.3390/ma11010024.

[10] S. v. Muniandy, C. S. Kan, S. C. Lim, and S. Radiman, "Fractal analysis of lyotropic lamellar liquid crystal textures," Physica A: Statistical Mechanics and its Applications, vol. 323, 2003, doi: 10.1016/S0378-4371(03)00026-8.

[11] A. Smith and V. M. Zavala, "The Euler characteristic: A general topological descriptor for complex data," Computers and Chemical Engineering, vol. 154, 2021, doi: 10.1016/j.compchemeng.2021.107463.

[12] I. Membrillo Solis et al., "Tracking the time evolution of soft matter systems via topological structural heterogeneity," Communications Materials 2022 3:1, vol. 3, no. 1, pp. 1–11, Jan. 2022, doi: 10.1038/s43246-021-00223-1.

[13] P. Pranav et al., "Topology and geometry of Gaussian random fields I: On Betti numbers, Euler characteristic, and Minkowski functionals," Monthly Notices of the Royal Astronomical Society, vol. 485, no. 3, 2019, doi: 10.1093/mnras/stz541.

[14] P. Bubenik, "Statistical topological data analysis using persistence landscapes," Journal of Machine Learning Research, vol. 16, 2015.

656c Physics-Informed Deep Neural Networks for Predicting Fluid Flow in Complex Porous Materials

Serveh Kamrava, Muhammad Sahimi

Fluid flow in heterogeneous porous media arises in many systems, from biological tissues to composite materials, soil, wood, and paper. With advances in instrumentations, high-resolution images of porous media can be obtained and used directly in the simulation of fluid flow. The computations are, however, highly intensive. Although machine learning (ML) algorithms have been used for predicting flow properties of porous media, they lack a rigorous, physics-based foundation and rely on correlations. We introduce an ML approach that incorporates mass conservation and the Navier–Stokes equations in its learning process. By training the algorithm to relatively limited data obtained from the solutions of the equations over a time interval, we show that the approach provides highly accurate predictions for the flow properties of porous media at all other times and spatial locations, while reducing the computation time.

656d Simulation of Neutron Dark Field Interferometry Data in Hierarchical Materials Using Small Angle Scattering Models

Caitlyn Wolf, Youngju Kim, Anis Ben Said, Sarah Robinson, Ryan P. Murphy, M. Cyrus Daugherty, Michael Huber, David L. Jacobson, Jacob LaManna, Nikolai Klimov, Paul Kienzle, Peter Bajcsy, Daniel S. Hussey, Katie Weigandt

Hierarchical materials can be found across many fields, including electrodes for alternative energy, pharmacology, biology, colloidal science, geology, construction, additive manufacturing, polymer science, and more. While small-angle neutron scattering (SANS/USANS) is a useful measurement tool for characterizing the structure in both hard and soft matter at length scales between 1 nm and 10 µm, it only provides a beam-averaged view of the structure, making the study of inherently heterogeneous materials difficult and limiting our understanding of the structure-function relationship that can span many length scales. In response, a new neutron far field interferometer (dubbed "INFER") is currently under development at the National Institute of Standards and Technology (NIST) that will enable the collection of spatially resolved, in three dimensions, structural information at the same length scales as SANS and USANS. This instrument will generate tomographic reconstructions with voxels on the order of 50 µm that each captures the local structure of the sample through the dark field intensity, which is related to the small angle scattering intensity through a single Hankel transformation. To analyze the large amounts of data expected with this approach (~ $10^5 - 10^6$ correlograms or a terabyte of raw data per day), machine learning and data science approaches will be critical to segmenting our sample into structurally-similar regions of interest. In this work, we discuss our recent progress in developing physics based INFER data simulation tools that enable us to generate the large amounts of training data required for these models during this time of instrument development. We make use of the existing libraries of form factors and structure factors in SasView that model a wide range of structural systems and materials in the SANS/USANS space and convert the data into the INFER space using the Hankel transformation. Moreover, this approach allows us to use our understanding of structural systems in the SANS or Fourier space to understand how structures appear in the INFER or correlation space and further define the applications and limitations of our instrument.

656e Developing an ML Algorithm to Predict the Aqueous Solubility of Polymers and Organic Compounds

Arash Tayyebi, Ali Alshami

In this work, we develop a machine learning (ML) algorithm for predicting the aqueous solubility of polymers and organic compounds. The solubility of components in various solvents can be nominated as an essential value for understanding the physical properties of various substances. Better and more reliable equilibrium solubility can be determined without the need for time-consuming experimental and pilot studies using an accurate machine learning algorithm by transforming conventional groups into a sequence of bits that computers can recognize. Conventional groups play an important role in calculating the solubility values, and ML can be a game-changer as a contribution method to calculate these measures. To determine conventional groups as the input data in ML method, structure-related information, such as atoms, rings, bonds and functional groups are required. In order to compare molecular structures, an abstractions called molecular fingerprints or molecular descriptors should be used. In this study, we trained an ML algorithm by using chemical descriptors and fingerprints methods. Molecular descriptors are used to illustrate a molecule's physical, chemical, or topological characteristics. We used seventeen descriptors in this study to predict aqueous solubility and compare their impacts on outputs results.

Molecular fingerprints used in this study is categorized in two shape: path-based and circular fingerprint. Topological or path-based fingerprints include combinations of atom types and paths between various atom types. In this type of fingerprint, fragments of the molecule are generated by following a path up to a certain number of bonds within the molecule. Path-based fingerprints hash all branched and linear molecular subgraphs up to a particular size by combining atom types, the atomic number, and aromaticity state with bond types. The Daylight fingerprint is the most well-known example of path-based fingerprints, and the RDKit fingerprint is a relative of the well-known Daylight fingerprint. In this study, a maximum path length of five (RDK5) was used.

Circular fingerprints are generated by considering the "circular" environment of each atom up to a given "radius" or "diameter" from the central atom. Morgan fingerprint, also known as extended-connectivity fingerprints ECFPs, is the most popular circular fingerprint which perceives the presence of specific circular substructures around each atom in a molecule. ECFPs are a method to identify identical molecules that have different atom numberings by representing the number of heavy-atom neighbors, the number of hydrogens, the isotope, and ring information. ECFPs have different types based on selecting different maximum bond lengths or diameters of the circular atom neighborhood where the digit at the end shows the maximum diameter value employed to generate the fingerprint. In this study, a circular fingerprint with a diameter 4 and 6, i.e., ECFP4 and ECFP6 were used.

We train ~2000 organic and polymeric compounds using the Random Forest (RFs) model as the regressors with the average R^2 test values around 0.91 and 0.81 and 0.82, respectively for molecular descriptors, path-based and circular fingerprints. The most important features of each method and their impact on the aqueous solubility were investigated in this study.

656g Machine Learning for the Discovery of Molecular Recognition Based on Single-Walled Carbon Nanotube Corona-Phases

Xun Gong, Nicholas J Renegar, Retsef Levi, Michael Strano

The nanoparticle corona phase (CP) offers a unique materials design space for constructs capable of molecular recognition (MR) for sensing applications. Single-walled carbon nanotube (SWCNT) CPs have the additional ability to transduce MR through its band gap photoluminescence (PL). DNA oligonucleotides are well-known to disperse SWCNTs through forming CPs and can be manufactured with molecular precision. Nevertheless, no generalized scheme exists for the de novo prediction of SWCNT MR based on these CPs due to their sequence-dependent three-dimensional complexity. This work generated the largest DNA-SWCNT PL response library of 1408 elements and leveraged machine learning (ML) techniques to understand the DNA sequence dependence of MR. Both local features (LFs) and high-level features (HLFs) of the DNA sequences were utilized as model inputs. Out-of-sample analysis of our ML model showed significant correlations between model predictions and actual sensor responses for 6 out of 8 experimental conditions. Different HLF combinations were found to be correlated with sensor responses for each analyte, offering mechanistically differentiable design parameters for these systems. Furthermore, models utilizing both LFs and HLFs show improvement over that with HLFs alone, demonstrating that DNA-SWCNT CP engineering is more complex than simply specifying molecular properties. Taken as a whole, this work details the feasibility and utility of a ML-guided approach for nanoparticle CP engineering with relatively few experiments within a high-dimensional design space.

656h Accelerated Discovery of Novel Ionic Liquid Cations Using a Continuous Latent Space Representation of Chemical Space

Pratik Dhakal, Jindal Shah

Ionic Liquids (ILs) have generated tremendous interest in the research community over the years due to several unique and desirable properties compared to conventional solvents. However, its slow transport properties and high material costs have limited its commercial application development. Due to the vastness of the ionic liquid chemical space, narrowing down the potential candidates using experiment or atomistic simulation with faster dynamics, high stability, and low material cost has proven to be difficult. Machine learning (ML) algorithms, on the other hand, could serve as an alternative tool for accelerated material discovery with desirable properties. This paper employs a generative-based deep machine learning method to discover new cations in the ionic liquid chemical space, thereby increasing the number of potential cation candidates for battery applications. According to our findings, some of the newly discovered cations belong to known cation families, whereas most of the cyclic/aromatic cations occupy a new region of chemical space. To assess their suitability as potential electrolytes, we calculate the electrochemical stability of these cations using DFT calculations. Few of the cations have very high electrochemical stability, making them ideal for battery applications.

656i Context-Aware Representations from Deep Learning for Antibody Design

Sai Pooja Mahajan, Jeffrey J. Gray

Protein-protein interactions drive most protein function including antigen recognition and neutralization, cell signaling, catalysis, etc. Designing proteins and antibodies to bind protein targets of interest such as the spike protein of the SARS-Cov2 or a protein receptor implicated in a disease is an important task in developing therapeutics and diagnostic tools. We formulate the protein interface design problem as the task of predicting the interfacial sequence of a protein binder given the sequence and structural context of the interface. For this task, we developed an equivariant graph neural network that learns residue-level representations at protein interfaces. Since the number of protein interfaces with solved structures is limited, we train our model to learn from all available structures and fine-tune on protein and antibody-antigen interfaces. The model generalizes residue representations from proteins to protein-protein and antibody-antigen interfaces learning context-aware representations. On a test set of 400 protein interfaces (non-redundant from training sets), our model recovers over 38% of the native interface sequence with an average maximum recovery of 66.7% per target. We further fine-tune the model on antibody-antigen interfaces. On a test set of over 70 antibody-antigen interfaces including single-domain antibodies (non-redundant from training sets), the model recovers 37.8% of the native interface with an average maximum recovery of 72.3% per target. Furthermore, for heavy chain hypervariable loops, H1, H2, H3, the model recovers 46.5%, 48.9% and 39.3% respectively. Our model presents a fast and context-aware approach to design proteins and antibodies in the context of their binding partners.

482a Mechanochemical Degradation Pathways of Protective Oxide Surfaces: Development of an Ab-Initio Informed Multiscale Corrosion Model

Jeremy Scher, Tae Wook Heo, Yue Hao, Matthew Kroonblawd

Protective surface coatings inhibit aluminum corrosion pathways by replacing the highly reactive metal surface with one that is more chemically stable. Chemical processes which erode these protective films render aluminum more susceptible to corrosion. Identifying the dominant chemical drivers for these erosion processes and the mechanisms by which they proceed can help to inform the design of more resilient and robust protective coatings. We use *ab initio* molecular dynamics to model the erosion of γ-alumina surfaces and calculate associated reaction barriers while subjecting the oxide surface to environmental stressors, including vacuum, pooled water, and mechanical stress. We find that the presence of water reduces erosion reaction barriers by over 60% with respect to vacuum. Connections with microstructure-aware micromechanics models show that localized shear stresses in these materials can develop, which can couple with an aqueous environment resulting in further erosion barrier reductions. These findings highlight the relevance of solvent and surface strain on aluminum corrosion and serve as a basis for predictive multiscale lifetime models.

This work was performed under the auspices of the U.S. Department of Energy by Lawrence Livermore National Laboratory under Contract DE-AC52-07NA27344.

482b Molecular Insights into NMR Relaxation of Gd(III)-Based Contrast Agents for MRI Applications

Thiago Jose Pinheiro Dos Santos, Arjun Valiya Parambathu, Dilip Asthagiri, Philip Singer, Walter Chapman

Nuclear magnetic resonance (NMR) relaxation is a powerful tool for probing matter non-destructively, with applications spanning oil recovery to material science to medicine. In this work, we seek to investigate NMR relaxation in the presence of gadolinium (III), which is a paramagnetic species used in MRI contrast agents. Gd(III) is a highly charged ion and here we model its interaction with liquid water using the AMOEBA polarizable forcefield. We carry out first-of-its-kind long time-scale molecular dynamics simulations of this system using GPU-acceleration. At frequencies relevant to MRI, our simulations of NMR relaxivity show good agreement with experiments, without any free parameters in the interpretation of the simulations. This is stark contrast to the extended Solomon-Bloembergen-Morgan (SBM) model which is the mainstay for interpreting relaxation in MRI.

A crucial ingredient in our approach is a way to model the relaxation for long times. We do this by decomposing the autocorrelation into contributions from different exponentially-relaxing modes. This allows us to model the response to long times and provides physical insights into the molecular-scale dynamics underlying relaxation. Being able to reconstruct NMR relaxation using molecular modes can prove transformative in practical applications. Having established our methods for Gd(III) at one temperature, we explore temperature effects and the role of chelating agents. In this talk, we will also discuss our efforts to develop forcefield parameters for the chelating agents using ab initio simulations.

482c Interface Force-Field (IFF) Parameterization of $Ti_3C_2T_X$ Mxenes

ISAAC Armstrong, Vikas Varshney, Hendrik Heinz

Since their discovery in 2011 by Gogotsi et al., two-dimensional (2D) MXene materials have attracted growing interest due to their multi-functional properties, such as high electric conductivity (for superior electromagnetic interference (EMI) shielding properties), and high in-plane mechanical strength (filler reinforcements for structural applications), among others. Given their inherent 2D nature, these materials possess low shear resistance in their pristine form, limiting their applicability in applications which require structural stability under working conditions. In this context, accurate modeling of shear behavior at nanoscale would illuminate desired surface features that can maximize their shear strength. This limitation necessitates development of a MXene force-field which can accurately model inter-flake interactions via atomistic molecular dynamics simulations, to provide insights on modulating shear and related surface characteristics of MXenes. In this work, we develop accurate and compatible force-field parameters for Ti3C2Tx, one of the most studied MXenes, with hydroxyl (-OH) and fluoride (-F) terminated surface functional groups as a part of the Interface Force Field (IFF). The IFF incorporates atomic charges and non-reactive chemical bonding as well as non-bonded van der Waals and electrostatic interactions. The model is validated against experimental reference data including lattice parameters obtained via X-Ray scattering, surface energies, contact angles, and in-plane/out-of-plane mechanical properties. Furthermore, a genetic-algorithm (GA) based machine learning framework interfaced with LAMMPS has been employed to refine and optimize the force-field parameters. The developed IFF model for MXenes can be used to simulate interfaces with organic and biological molecules, solvents, and inorganic materials without additional parameters. The model setup and parameters outlined in this work are easily extendable to other types of MXene materials.

482d Molecular Characterization of High Ionic Conductivity in Fluoroether Lithium Metal Battery Electrolytes

Yuxi Chen, Elizabeth Lee, Juan J. de Pablo

In hopes of designing and engineering next generation electrolytes for dendrite-free lithium metal batteries with sufficiently high conductivity, we studied the underlying Li-ion transport mechanisms in several novel fluoroether solvents. Here, we use all-atom molecular dynamics (MD) simulations to elucidate the properties of ion transport and solvation for several fluoroether chains with varying lengths of ether segments and the number of fluorinated terminal groups. We show that ionic conductivity can be tuned by changing the fraction of ether and fluorine contents. More specifically, we find that solvents with one fluorinated terminal group generally outperforms those with two fluorinated terminal groups with higher ionic conductivity and more stable solvation environment. We find that the solvent size determines the lithium-ion transport kinetics, solvation structures, and solvation energies. As the length of the fluoroether chain increases, lithium ions bind more strongly to the solvent. Consequently, the lithium-ion transport mechanism shifts from ion hopping between solvation sites in different fluoroether chains in short-chain solvents to the ion-solvent co-diffusion in long-chain solvents. Free energy calculations reveal that longer solvent molecules have lower solvation free energies, stabilizing lithium ions. Experimental measurements predict a surprising non-monotonic behavior of solvent size effect on the lithium-ion conductivity, where the maximum ionic conductivity occurs at the medium-length solvent chains, which agrees with our analysis. Achieving high ionic conductivity, therefore, requires a trade-off between increasing the solvent self-diffusivity by using shorter-length fluoroether chains and increasing the stability of local solvation shells by using longer-length chains. Our study demonstrates that this critical balance between high solvent self-diffusivity and low solvation energy is a key design principle for achieving lithium battery electrolyte solutions with high lithium-ion conductivity.

482e Ab Initio Study on the Reaction Kinetics of Ethylene Glycol Decomposition on Pt (111) and Pt_3sc (111).

Shedrack G. Akpe, Hyung Chul Ham

Rising global energy demand and rapid depletion of fossil fuel reserves are driving the search for alternative energy resources. Recently, there has been an increasing interest in using lignocellulose biomass as a sustainable feedstock for fuels and value-added chemicals, resulting in valuable fuels mostly with zero CO_2 emission. Ethylene glycol (EG) conversion is relevant for hydrogen and syngas production. Platinum is the most effective catalyst for this process due to its synergistic features of high dehydrogenation and high selectivity to C–C bond scission activities; however, it is economically intensive. The use of bimetallic catalysts for biomass feedstock refining is an intelligent technique because the interaction between metals can alter the surface properties of the catalyst. These modifications can greatly enhance catalytic activity, alter selectivity to the targeted products, and boost catalyst stability in the presence of biomass-derived impurities and harsher reaction conditions. In this study, we investigated the mechanism of EG decomposition for hydrogen production on Pt (111) and Pt_3Sc (111) from first-principles calculations. We reported the various barriers and reaction energies by exploring all the 63 elementary reactions involve and developed microkinetic models that elucidate the vapor phase kinetic data for the process.

Furthermore, we presented the minimum energy pathway (MEP) on Pt (111) and Pt_3Sc (111) with the aid of potential energy diagrams (PEDs). Finally, we compared the activities of both catalysts and demonstrated how Pt_3Sc (111) outperformed Pt (111) in terms of the MEP barriers. Our results revealed that on Pt (111) surface, EG dehydrogenation underwent a competitive pathway starting either at C–H or O–H bond scission as obtained from the almost similar Gibbs' free energy barrier of the first stage dehydrogenation process. However, on Pt_3Sc (111) surface, EG dehydrogenation conspicuously commenced with O–H bond scission with a very low barrier and proceeded until complete dehydrogenation was achieved.

The results of the dehydrogenation pathways on both surfaces are presented in Fig. 1 via Fig. 9, and the parameterization for the microkinetic studies is shown in Tables 3 and 4.

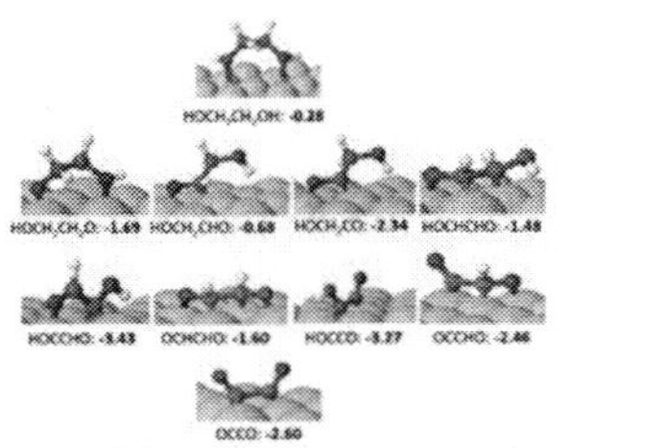

Fig.1. Intermediates Conformations of EG Dehydrogenation on Pt (111) starting at O–H scission

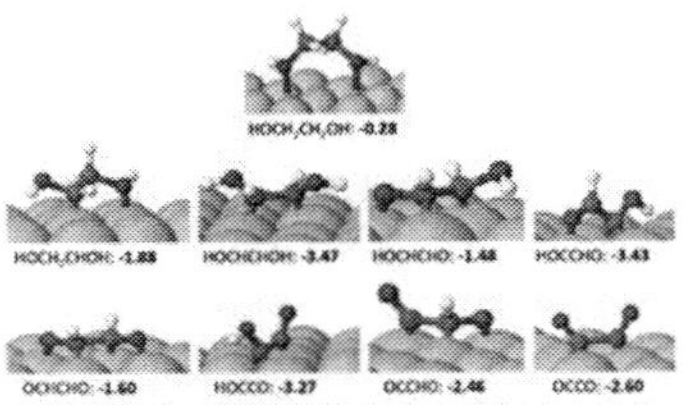

Fig.2. Intermediates Conformations of EG Dehydrogenation on Pt (111) starting at C–H scission

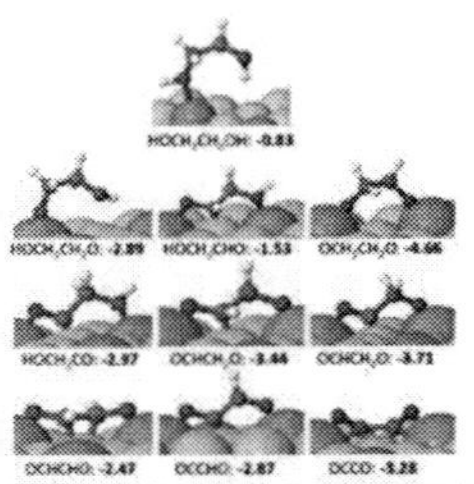

Fig. 3. Intermediates Conformations of EG Dehydrogenation on $Pt_3Sc(111)$ surface

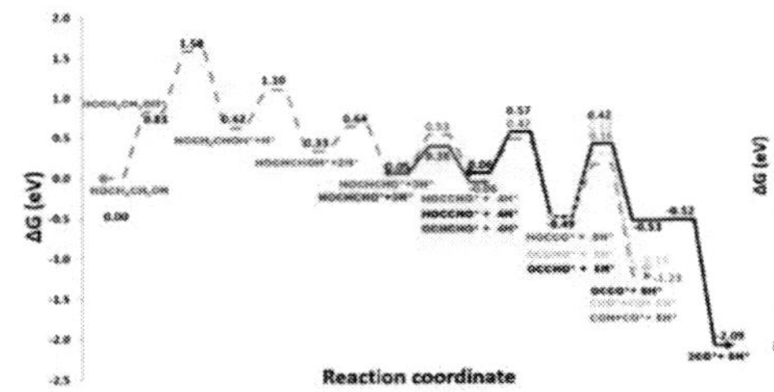

Fig 7. Potential energy diagram of EG complete dehydrogenation on Pt (111) starting at C–H bond scission.

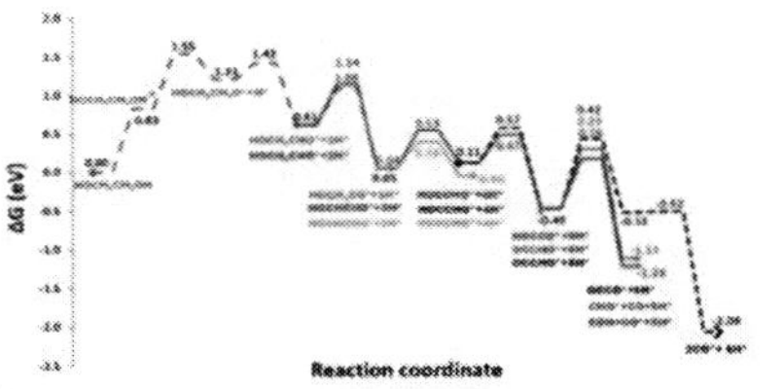

Fig 8. Potential energy diagram of EG complete dehydrogenation on Pt (111) starting at O–H bond scission.

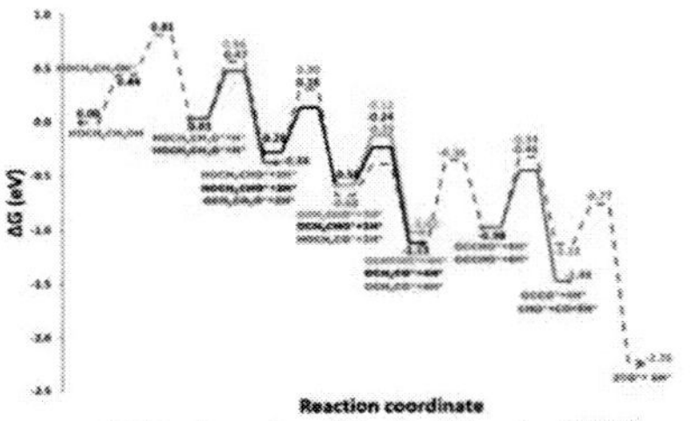

Fig 9. Potential energy diagram of EG complete dehydrogenation on Pt_3Sc (111).

482f Oxidation Rate and Leaching of Flame-Made Nanosilver By Reactive Molecular Dynamics

Eirini Goudeli, Diego Chaparro

Nanosilver is a nanomaterial commodity used commonly into consumer products, such as food containers, bandages, and cosmetics. In addition, silver nanoparticles show high antibacterial activity finding applications in biomedical products. However, there are concerns about the adverse effects of nanosilver on human health due to release of silver ions from the nanoparticle surface, as leaching is associated with nanosilver toxicity. Prior experimental studies (Sotiriou et al., 2012) indicated that the release of silver ions originates from one or two oxidized monolayers formed at the nanoparticle surface.

Here, the oxidation of silver nanoparticles made by flame spray pyrolysis is investigated at high temperature by reactive molecular dynamics. The oxidation rate equation is obtained *ab initio* as a function of temperature and particle coverage. Figure 1 shows the fraction of silver ions in nanosilver particles as function of particle size obtained by reactive molecular dynamics simulations (diamonds). The results are in excellent agreement with experiments of flame- and wet-made nanosilver (triangles) in aqueous suspensions. For comparison, the estimated silver ion fractions assuming one (circles) and two (squares) silver oxide layers are shown. Nanosilver particles with diameter of 3 – 4 nm exhibit a large fraction of silver ions, which are present in the form of silver oxide at the particle surface. This fraction corresponds to a double layer of silver oxide. Larger silver nanoparticles, having diameter of 7 – 8 nm, exhibit a smaller fraction of silver ions that corresponds to a single silver oxide layer, consistent with experiments (Sotiriou et al., 2012). In addition, the effect of particle size on the leaching rate of nanosilver in aqueous solutions is quantified to understand the mechanism of silver ion release, which is responsible for the nanosilver toxic manifestations. Minimising these release rates can enable the expanded use of silver nanoparticles and their nanocomposites as bioimaging materials or plasmonic biosensors.

References

Sotiriou, G. A., Meyer, A., Knijnenburg, J. T., Panke, S., andamp; Pratsinis, S. E. (2012). Quantifying the origin of released Ag+ ions from nanosilver. Langmuir, 28(45), 15929-15936.

482g Molecular Dynamic Insights on the Distinct Solvation Structures of Aromatic and Aliphatic Compounds in Monoethanolamine Based Deep Eutectic Solvents

Nikhil Kumar, Tamal Banerjee

Deep eutectic solvents (DESs) are developing as an alternate medium for aromatic extraction, especially benzene and thiophene from aliphatic hydrocarbon mixtures. In this work, molecular dynamics (MD) simulations were first used to investigate the solvation structure of benzene, thiophene, and n-hexane in monoethanolamine-based DESs. It reveals the liquid structures in the adjacent neighbour shells, which is a function of electron-withdrawing sulphur attached to thiophene and the π-electron cloud of benzene. The intermolecular forces between aromatic, aliphatic, and DESs components are analysed in van der Waals and hydrogen bond interactions. The chloride ions serve as a charge carrier between choline and monoethanolamine precursors. The solvation of benzene, thiophene, and n-hexane in the deep eutectic solvent depends on volume expansion and minor solvent structural changes. The solvation shell structure and characteristics were investigated in tandem with the possibility of benzene and thiophene clustering. The 1H NMR and 2D 1H–1H-NOESY were used to investigate the intermolecular interactions between benzene, thiophene, and n-hexane with monoethanolamine-based solvents. It concludes that high-ordered DES1 is more inclined to higher solubility than lower-ordered ones with a higher molar ratio of monoethanolamine. The solvation was reduced because the entropy gain was not maximised in the lower ordered DESs.

181a Optiboost: A Method for Choosing a Safe and Efficient Boost for the Bond-Boost Method in Accelerated Molecular Dynamics Simulations with Hyperdynamics

Jianming Cui, Kristen Fichthorn

Accelerated molecular-dynamics (MD) simulations based on hyperdynamics (HD) can significantly improve the efficiency of MD simulations of condensed-phase systems that evolve via rare events. However, such simulations are not easy to apply since the optimal boosts are usually unknown. In this work, we developed a method called OptiBoost to adjust the value of ΔV_{max} (i.e., the boost) in HD simulations based on the bond-boost method. The OptiBoost method involves (1) Selecting a value of ΔV_{max} and running a series of short trajectories to obtain the probability $P(\Delta V_{max})$ that an event occurred; (2) Fitting $P(\Delta V_{max})$ to a sigmoid function and; (3) Delineating the boost range from the maximum in $P'(\Delta V_{max})$ and from the minimum in $P''(\Delta V_{max})$. We demonstrated the OptiBoost method in simulations on a cosine potential and applied it in three different systems involving Ag diffusion on Ag(100) in vacuum and in ethylene-glycol solvent. In all cases, OptiBoost was able to predict safe and effective values of the boost, indicating the OptiBoost protocol is an effective way to advance the applicability of HD simulations.

181b Coupling Replica Exchange with Backbone sampling Captures Conformational Changes on Protein-Protein Interfaces

Ameya Harmalkar, Jeffrey J. Gray

Protein-protein interactions (PPIs) are involved in almost all biological processes in human health and disease. Understanding the structure of a protein complex can reveal biological mechanisms and suggest intervention strategies. Since experimental techniques are often infeasible, docking provides an alternative to elucidate structures and guide molecular engineering. Recently, AlphaFold *(Jumper et al. 2021; Evans et al. 2021)* and other deep learning algorithms *(Baek et al. 2021)* have made a breakthrough in the field of protein sequence-to-structure prediction. Yet, modeling protein interfaces and predicting conformational changes, remains an unsolved challenge, especially when there is significant conformational change in one or both binding partners. To overcome these limitations, we introduce a new enhanced sampling algorithm, called Hamiltonian-Resolution replica exchange (H-ResEx). The H-ResEx algorithm improves canonical sampling over rugged, atomistic energy landscapes by swapping conformations between the coarse-grained (CG) and all-atom (AA) states. To capture large-scale conformational changes, we mimic induced-fit approach of protein binding and sample backbone moves on-the-fly. We demonstrate the performance of our method for docking protein targets on a benchmark set of 100 proteins with moderate to high flexibility (unbound to bound RMSD over 1.2 Å up to 8 Å). Moreover, our advanced simulation approach equips the computational advantages of coarse-graining and requires 300-500 CPU hours for a docking simulation (depending on protein sizes), which is efficient compared to molecular dynamics-based approaches. With this work, we show that a CG/AA exchange scheme with conformational sampling shows substantial promise towards quantitative and qualitative modeling of protein complexes and their dynamic interactions. The proposed algorithm paves the way towards challenging applications in enhanced sampling of biomolecular systems, employing coarse-grained models for capturing conformations without compromising on the accuracy of the all-atomistic models.

181c Molecular Modes from NMR Relaxation in Fluids: Going Beyond the Bpp Theory

Arjun Valiya Parambathu, Philip Singer, Thiago Jose Pinheiro Dos Santos, George J. Hirasaki, Walter Chapman, Dilip Asthagiri

Nuclear Magnetic Resonance (NMR) relaxation is a phenomenon by which nuclei with a non-zero spin return to equilibrium in the presence of an applied static magnetic field after it has been excited by an applied radio-frequency magnetic field. It is characterized by two time constants, T1 and T2, corresponding to relaxation in the longitudinal and transverse directions, respectively. T1 and T2 are dependent on the equilibrium structure and dynamics of the system, providing an avenue to probe matter non-destructively. For close to 70 years, the interpretation of T1 and T2 in liquids has relied on the Bloembergen-Purcell-Pound(BPP) theory, which made rather severe assumptions about the fluid structure and dynamics, such as treating molecules as hard spheres and the magnetic dipoles as freely rotating, features that ultimately justify a mono-exponential decay of magnetic dipole-dipole autocorrelations.

Analysis of magnetic dipole-dipole autocorrelations from molecular simulations shows that even for simple alkanes and water, the mono-exponential assumption is not met. We show that the T1 relaxation in the slow-motion regime of crude oils, hydrocarbon viscosity standards, and glycerol all significantly depart from the BPP theory. The NMR relaxation autocorrelation exponents are distinctly different and reveal the existence of a distribution of correlation times within the autocorrelation. This distribution appears to be multimodal on the logarithm of the autocorrelation times. The modes correspond to the molecular structure and environment studied, hence termed molecular modes.

I will present our findings on NMR relaxation from simulations in bulk fluids, fluids under confinement, and water in the presence of paramagnetic ions. I will show that the assumptions underlying BPP theory are invalid, a finding that has far reaching implications for how one interprets NMR relaxation phenomena in condensed media. I will also show the potential for using molecular modes distribution to enhance the interpretation of NMR relaxation.

181d Unbiased Coarse-Grained Monte Carlo Simulation Using SAXS-Data for Identification of Self-Assembled Nanostructures

Silabrata Pahari, Shuhao Liu, Mustafa Akbulut, Joseph Kwon

Over the last decade, there has been an increasing interest in industry and academia to develop complex fluids that can change their rheological properties reversibly in response to the changes in solution conditions [1]. These reversible fluids find great applications in the pharmaceutical, oil and gas, and specialty chemical industry sectors [2]. Specifically, the reversible nature of these fluids is attained via the dynamic nanostructure of self-assembled amphiphiles which are the building blocks of these complex fluids [3]. By carrying out deeper characterization and detailed free energy calculations, it was revealed that these self-assemblies have unique nanostructures and identifying these nanostructures is necessary as important properties of the complex fluids, like their viscoelastic properties, are strongly correlated with the geometry and dimensions of these nanostructures [4]. Therefore, it has been widely studied to develop effective characterization methods that can help us to identify the nanostructures present in such complex fluids [4]. However, it has been difficult to develop a general methodology that can help us to determine the detailed geometry and dimensions of the nanostructures present in the self-assemblies.

It is difficult to characterize the complex self-assembles because these techniques employ artificial treatments like dehydration, which can often lead to inaccurate imaging results due to significant changes in the morphology of the complex fluids [5]. To address these limitations, methods like cryogenic transmission electron microscopy (cryo-TEM) were utilized. However, sample preparation is challenging in cryo-TEM; additionally, rapid cooling in cryo-TEM may perturb the amphiphile conformation [6]. Furthermore, when scattering methods like small-angle X-ray scattering techniques (SAXS) are used to obtain the high-resolution three-dimensional description of the nanostructures in dispersed phase, there arises an inherent challenge associated with the degeneracy resulting from the reconstruction of three-dimensional nanostructures from one-dimensional SAXS scattering profiles [7]. Henceforth, solely utilizing SAXS scattering profiles, only simple structural features such as radius of gyration and maximum diameter of the nanostructures can be extracted. Recently SAXS-guided molecular dynamics (MD) simulations have been extensively carried out to characterize to native structures of biopolymers that are difficult to obtain via conventional experimental characterization techniques [8]. In these simulations the total Hamiltonian of the system is biased by adding a potential term which is a function of the discrepancy value between the SAXS profiles obtained experimentally and SAXS profiles calculated from the molecular coordinates in the MD simulations. Although this method has been successful in characterizing native structures of biopolymers they fail to characterize complex nanostructures because it requires the calculation of the *ab-initio* SAXS profiles during the MD simulations which becomes computationally intractable in the cases of large-scale nanostructures.

To address the limitations of the existing experimental and simulation techniques in obtaining a detailed three-dimensional description of complex self-assembled nanostructures, a SAXS-guided unbiased coarse-grained Monte Carlo (CGMC) simulation methodology is proposed. In this method, two metrics are utilized to identify the detailed molecular description of the complex self-assemblies formed by the amphiphiles. The first metric used is the SAXS discrepancy values, and the second metric is the total potential energy calculated from the CGMC simulations. In this proposed method, the Hamiltonian of the system is not biased, so the energy calculated from coarse-grained Monte Carlo simulations is accurate and can be utilized as an additional metric for structure identification. In this method, first, a library of nanostructures is generated, and then CGMC simulations are seeded from these structures. From the results of the CGMC simulations, structures with the least SAXS discrepancy values and the lowest energies are screened for the next round of sampling. Subsequently, from the screened structures, new simulations are further seeded, and the process is continued till no improvement in the SAXS discrepancy values and the total potential energy of the system is obtained. The proposed methodology is applied to the system of complex self-assembly called the dynamic binary complexes (DBC), which are formed by the self-assembly of OAPB amphiphiles and DTA ions. Henceforth, the SAXS-guided unbiased CGMC simulations could identify the nanostructures present in the DBCs. Furthermore, the dimensions of the resulting nanostructures were validated with scanning electron microscopy (SEM) images. Overall, we resolved the fundamental ambiguity resulting from reconstructing nanostructures from SAXS scattering profiles with the proposed method.

References.

[1] Dexter, A. F., Malcolm, A. S., andamp; Middelberg, A. P. (2006). Reversible active switching of the mechanical properties of a peptide film at a fluid–fluid interface. *Nature Materials*, *5*(6), 502-506.

[2] Öttinger, H. C. (1998). General projection operator formalism for the dynamics and thermodynamics of complex fluids. *Physical Review E*, *57*(2), 1416.

[3] Butler, P. (1999). Shear induced structures and transformations in complex fluids. *Current Opinion in Colloid andamp; Interface Science*, *4*(3), 214-221.

[4] Bhat, B., Liu, S., Lin, Y. T., Sentmanat, M. L., Kwon, J., andamp; Akbulut, M. (2021). Supramolecular dynamic binary complexes with pH and salt-responsive properties for use in unconventional reservoirs. *PloS one*, *16*(12), e0260786.

[5] Liu, S., Lin, Y. T., Bhat, B., Kuan, K. Y., Kwon, J. S. I., andamp; Akbulut, M. (2021). pH-responsive viscoelastic supramolecular viscosifiers based on dynamic complexation of zwitterionic octadecylamidopropyl betaine and triamine for hydraulic fracturing applications. *RSC Advances*, *11*(37), 22517-22529.

[6] Hsieh, C., Schmelzer, T., Kishchenko, G., Wagenknecht, T., andamp; Marko, M. (2014). Practical workflow for cryo focused-ion-beam milling of tissues and cells for cryo-TEM tomography. *Journal of structural biology*, *185*(1), 32-41..

[7] Livet, F., Bley, F., Ehrburger-Dolle, F., Geissler, E., Lebolloc'h, D., andamp; Schulli, T. (2003). Ultra small-angle X-ray scattering: conditions, limits and results in speckle experiments. *Journal of applied crystallography*, *36*(3), 774-777.

[8] Kimanius, D., Pettersson, I., Schluckebier, G., Lindahl, E., andamp; Andersson, M. (2015). SAXS-guided metadynamics. *Journal of chemical theory and computation*, *11*(7), 3491-3498.

181e Equilibrium Adsorption Morphologies of Surfactants at Metal-Water Interfaces Studied Using a Novel Free Energy Sampling Methodology in Molecular Simulations

Sumit Sharma, Himanshu Singh

Adsorption of surfactants is a facile way of adjusting interfacial properties of metals, which has applications in electrochemistry, corrosion inhibition, heterogeneous catalysis and synthesis of anisotropic metal nanoparticles. Surfactant molecules are understood to adsorb in densely packed organized morphologies at the metal-water interfaces. Traditional molecular simulation methods as well as existing free energy sampling methodologies are inefficient in studying the formation of these morphologies due to the long timescales and large free energy barriers involved in the process. In this work, I will discuss a new umbrella sampling-based simulation methodology that our research group has developed that allows the study of these high-density morphologies. Using this methodology, we have investigated the adsorption behavior of different cationic, anionic and charge-neutral surfactants and surfactant mixtures. The role of molecular properties is governing the equilibrium adsorption morphologies will be discussed.

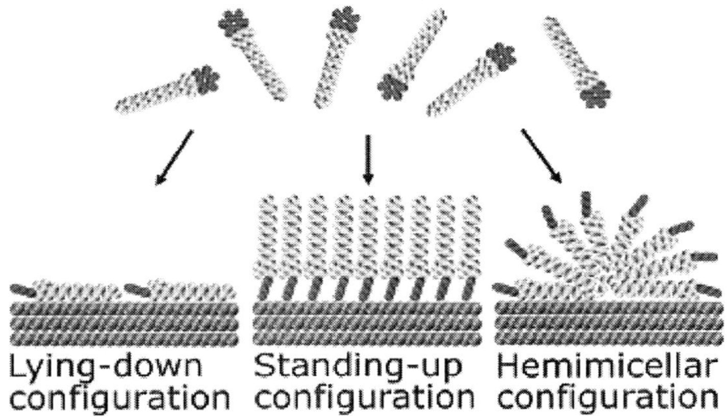

181f A Molecular Dynamics Simulation Study of the Kapitza Heat Transfer Resistance

Sebastian Schmitt, Simon Stephan, Hans Hasse

The heat transfer across a solid-fluid interface was systematically investigated by molecular dynamics simulations of a model system [1]. The studied system consists of a fluid confined between two parallel planar walls. There is no convection. Both the fluid and the solid were modeled with the Lennard-Jones truncated and shifted potential. The following parameters were varied systematically: channel height, temperatures of the wall and the fluid, density of the fluid, strength of the solid-fluid and the solid-solid interaction, and the mass of the solid particles. From the simulation results, numbers for the Kapitza length L_K were obtained, which characterizes the heat transfer resistance at the solid-fluid interface. A correlation of the results for the Kapitza length L_K as a function of the studied variables was developed. Moreover, a dimensionless number is introduced, the Kapitza interface number *Ki*, that describes the Kapitza effect in the stagnant fluid and is zero in the absence of the Kapitza effect. It is discussed, how the results from the present work can be used to assess the influence of the Kapitza effect in practical applications and how they can be applied to cases in which convection plays a role.

[1] S. Schmitt, T. Vo, M.P. Lautenschlaeger, S. Stephan, H. Hasse, *Mol. Phys. (2022), in press.*

181g Molecular Modeling of Alcohol Effects in Nonionic Surfactant Micelles with Density Functional Theory

Jinxin Lu, Walter Chapman

Surfactant micelles have been of interest for the past several decades because of its importance in both scientific studies and industrial processes. Among the large number of different types of additives that have been tested and studied, alcohol is among the most frequently used. It is known that alcohol can act as co-solvent by altering solvent properties, and it can also act as co-surfactant by inserting themselves into the palisade layer of the micelle and co-aggregate with surfactants. In most cases alcohol act dualistically and the overall effect depends on alcohol carbon chain length. Several thermodynamic models have been developed to predict the critical micelle concentration (CMC) and size distribution of micelles, but few of them is able to fully capture the physics at molecular level and predict the dualistic role of alcohol additives.

We study the micelle formation of nonionic poly (ethylene oxide) alkyl ether surfactant using interfacial Statistical Associating Fluid Theory (iSAFT), a classical density functional theory. The model is used to predict the CMC, aggregation number, and distribution of solute into a single micelle. We also study the effect of adding linear 1-alcohols into the surfactant solution. Our model is able to capture both cosolvent and cosurfactant effects and effectively predict the trends of alcohol ranging from short-chain to long-chain in qualitative agreement with experimental data. The theory can provide detailed density profiles of each component in the system to illustrate the locus of alcohol inside a micelle. We observe that all alcohols studied are present in the palisade layer in different amounts. Partition coefficient, aggregation number, micelle size, and total number of surfactant and alcohol molecules are also predicted and discussed. Furthermore, results from different surfactant architecture are compared and analyzed.

181h Extracting Solid-Melt Interfacial Free Energy and Anisotropy Strength of Al-Cu Alloy Using Molecular Dynamics Simulations

Amrutdyuti Swamy, Pabitra Choudhury, Daniel Dolce

Laser Melt Deposition is one of the growing additive manufacturing techniques employed in industries for creating new and improved materials. During this process, rapid solidification of the melt has been known to form cracks for an Al-Cu system. These cracks impede the overall strength of the material. Understanding the changes in interfacial free energy can help throw light on crystal growth andamp; crack propagation during these rapid solidification operations. We have tried to analyze the role of undercooling and solute composition on interfacial free energy for an Al-Cu system using molecular dynamics and the capillary fluctuation method.

An Al-Cu phase diagram was first generated for the interatomic potential in use. Crystal-melt interfaces were then created for different undercooling regimes using molecular dynamics, equipped with the solute composition information from the phase diagram. The equilibrium crystal-melt interfaces are then analyzed using the capillary fluctuation method to obtain the interfacial free energy and its anisotropy parameters. Our results show that the interfacial energy for the binary Al-Cu system does not have a linear dependence with solute composition or the amount of undercooling. The results also correctly predict the transition of dendrite growth orientations from (100) to (110) with Cu doping.

Acknowledgments: The work is supported by ARL Grant No. W911NF-2020032 and used the Extreme Science and Engineering Discovery Environment (XSEDE) TACC at the stampede2 through allocation [TGDMR140131]. This work utilized resources from the University of Colorado Boulder Research Computing Group, which is supported by the National Science Foundation (awards ACI-1532235 and ACI-1532236), the University of Colorado Boulder, and Colorado.

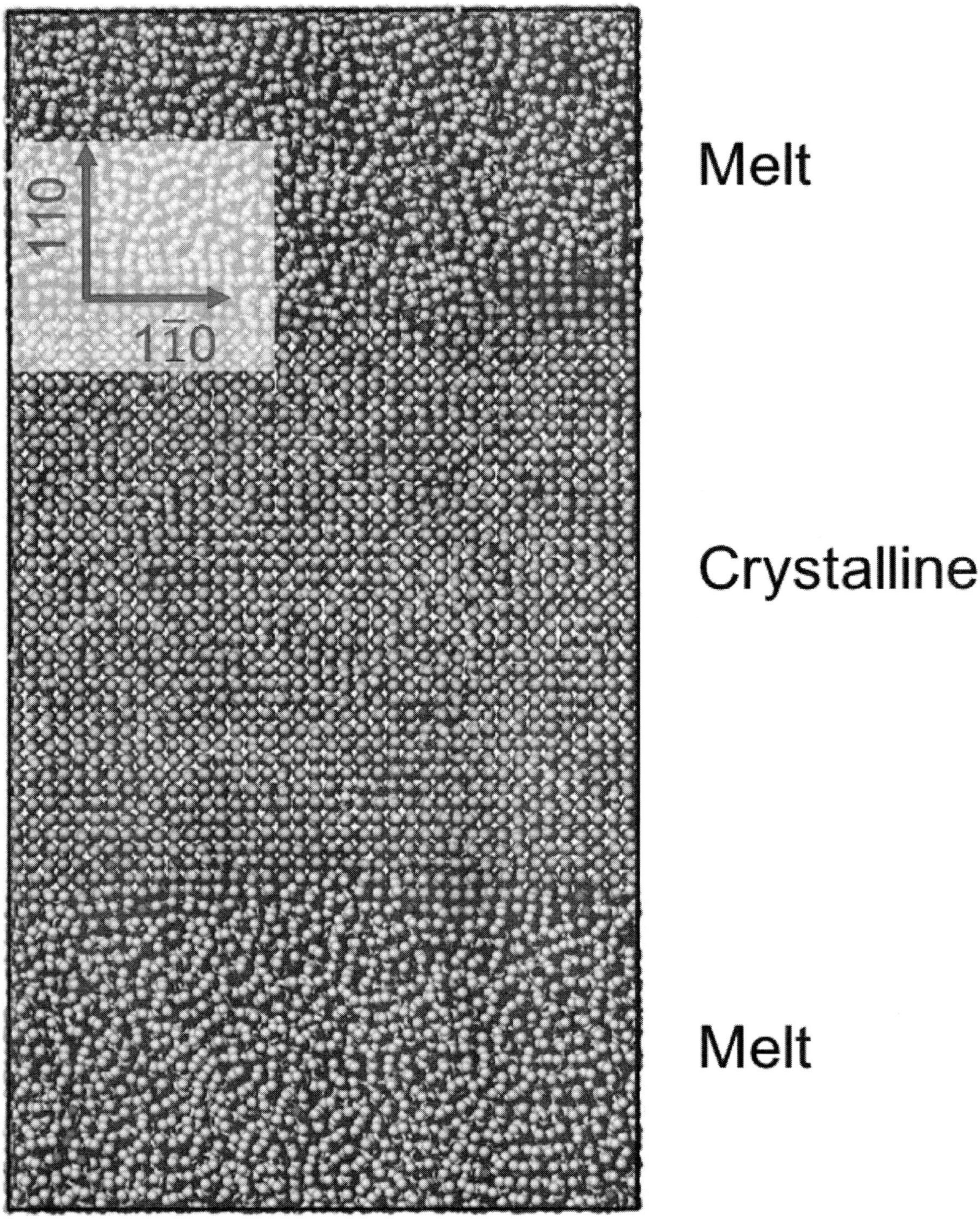
110
1$\bar{1}$0
Melt
Crystalline
Melt

181i Molecular Dynamics Simulation of O2 Transport in Nafion

Nicholas Tiwari

Polymer electrolyte membrane fuel cells (PEMFCs) are a viable energy source for small scale power generation. However, the power density of modern PEMFCs is still too low to provide adequate performance. Poor O_2 transport in the ionomer layer of the PEMFC adversely impacts power density. In this research, we use molecular dynamics simulations to explore the effect of ionomer chemistry on O_2 transport. Bulk calculations are used to quantify O_2 diffusivity, while thin-film calculations of ionomer on platinum catalyst are used to quantify O_2 solubility. Simulations show that the addition of bulky groups to ionomer backbones, and shorter ionomer side chains result in improved O_2 permeability. This is attributable to reduced ionomer crystallinity in the bulk phase and suppression of the formation of a dense ionomer layer in interfacial systems. These observations are corroborated by experiments.

181j Coarse-Grained Molecular Dynamics Investigation of Adsorption of Type IV Pili Proteins Onto Graphene-Cu(111) and Defective Graphene-Cu(111) Interfaces

Sourav Verma, Marina Davidson, Kenneth Benjamin

Despite their prevalence in natural and engineered systems, the exact molecular mechanisms and forces controlling biofilm formation and adhesion are relatively unknown. While most modeling studies to date have focused more on continuum or macroscopic film behavior, investigations at the molecular level are needed, too. Moreover, surface engineering of systems to prevent microbial adhesion and biofilm formation has become a key challenge for both biomedical and materials/corrosion communities.

Recent work has hypothesized that biofilm formation and adhesion may be related to the adsorption of key, early protein molecules (such as type IV pili proteins) and exopolysaccharides (EPS) included in conditioning films (CFs) to metal surfaces preceding the attachment of microbes. Reducing this early biofilm layer attachment to abiotic surfaces has shown great promise in mitigating biofilm formation, and in turn reducing microbial induced corrosion of metal surfaces. Graphene and graphene derivatives are potential candidates to be used as biofilm inhibiting coatings for metal surfaces due to their potential antibacterial properties.

In this study, we use coarse-grained molecular dynamics simulation to investigate the dynamic process of adsorption of Type IV pili (T4P) onto graphene and graphene-modified Copper slab (moire' superlattice Gr-Cu{111}) and its conformation change after adsorption on the surface(s), as a key step towards understanding the fundamental molecular level interactions that occur at the biofilm-surface interface. All the simulations were conducted in vacuum and in the presence of coarse-grained water as solvent to deduce the effect of solvation on the adsorption behavior.

Specifically, the molecular dynamics simulations are conducted using the LAMMPS molecular dynamics simulation software package. The coarse-grained MARTINI force field was used to capture the structural and thermodynamic properties of the key early conditioning film biomolecules (peptides and proteins) including Type IV pili. Graphene, it's derivatives (defective/pristine/multi-layer graphene) and Gr-Cu{111} substrate were modelled using a combination of the Adaptive Intermolecular Reactive Empirical Bond Order (AIREBO) potential to describe the C-C interactions, Embedded Atom Method (EAM) potential to describe Cu-Cu atoms interactions, and Lennard-Jones 6–12 potential to describe only nonbonded Cu-C interactions. The adsorption energies and the binding free energies of proteins on graphene-coated-Cu{111} and pristine/multi-layer/defective graphene modified surfaces are evaluated to study the adsorption phenomena. The binding free energies are computed using the metadynamics technique.

The results of this molecular-level study should aid in developing a larger, fundamental understanding of the interaction, adsorption, and adhesion of proteins and microbes to two-dimensional surfaces (with and without defects) and metal substrates, such as found in industrial and biomedical applications.

585a Structure Enumeration Algorithm for Silicone Mq-Resins

Steven G. Arturo, Thomas D. Bekemeier, Don Eldred, Wei Gao, James H. Wade, Tianlan Zhang

Silicone MQ-resins are key materials for rheology modification of polydimethylsiloxane fluids and as intermediates to be post-functionalized into high performance materials. The polymers within silicone MQ-resins are products of polycondensation from tetravalent Q-units (a central silicon atom with four bonded oxygen atoms) and monovalent M-units (a central silicon atom with one bonded oxygen atom and three bonded methyl groups). An effort to enumerate all possible molecular structures with minimal assumptions and minimal human intervention is made for comparison with published work on MQ-resin structures. Structures with up to 16 Q-units are found to be more globular in nature than those presented in previously published research.

585c Force Fields for the Prediction of Transport Properties of Lubricants at Extreme Conditions

Sebastian Schmitt, Simon Stephan, Hans Hasse

Lubricants are exposed to extreme conditions in tribological contacts; e.g. pressures in rolling bearings can be as high as 4,000 MPa. For such conditions, no experimental data on the density and the transport properties of the lubricants are available. However, information of these properties is needed for the design of the lubricated contacts. The currently applied strategy to solve this problem is to extrapolate experimental data obtained at mild conditions, which is obviously risky. Therefore, in the present work molecular models of lubricants were investigated regarding their ability to predict data that is relevant for the design of lubricated contacts at extreme pressures. Five different hydrocarbons were studied: three linear alkanes (decane, eicosane and triacontane) and two branched isomers of triacontane: squalane and 1-decene-trimer. The latter is a common component of poly-α-olefine (PAO) based oils. Molecular models from the literature were selected and compared to experimental data of different thermophysical properties in the range where such data were available. Molecular models of different types were considered: all-atom, united-atom, coarse-grained, as well as reactive force fields. They cover a wide range of complexity of modeling molecular interactions. Equilibrium molecular dynamics simulations were carried out to obtain data for the density, shear viscosity, thermal conductivity and the self-diffusion coefficient. Some of the studied force fields yield remarkably good predictions. The study enables the choice of appropriate force fields to determine transport properties of long hydrocarbons.

585d Understanding Electronic Properties of 1-n-Alkyl-3-Methylimidazolium Chloride Ionic Liquids with Iron Porphyrin

Sudip Kumar Das, Jindal Shah

Ionic liquids have garnered considerable attentions as potential replacements for volatile organic solvents in industrial applications due to their unique properties such as low vapor pressure, low flammability, and design flexibility. However, they cannot be assumed to be environmentally friendly, as many ionic liquids have been shown to be toxic to microorganisms. It has also been demonstrated that several ionic liquids are recalcitrant to biodegradation. However some ionic liquids do exhibit favorable biodegradable characteristics. Cytochrome P450 bearing iron porphyrin in their catalytic pocket has been proposed as a potential enzyme in the ionic liquid biodegradation. However, it is not clear how ionic liquids interact with FeP in the binding pocket. In this presentation, we investigate the interaction of ionic liquids 1-n-alkyl-3-methylimidazolium $[C_n mim]^+Cl$ (n = 2, 4, 6, 8, and 10) with FeP using electronic structure calculations. As these systems are characterized by a large number of degree of freedom, we will demonstrate how a thorough exploration of the available conformational space is required to search for an energy minimum. We will characterize the final optimized geometry with a coordinate system that is affixed to FeP so that the structures across the homologous series can be compared. Results of electronic properties such as the highest occupied molecular orbital energy and the lowest occupied molecular orbital energy will be presented. We will also calculate the reduction potential of the optimized structures as it represents the second step in the overall catalytic cycle that leads to the biodegradation of imidazolium-based ionic liquids.

585e Applying Deep Learning to Accelerate Molecular Dynamics Simulation-Based Structural Properties Prediction for Biomolecules.

Pin-Kuang Lai

Molecular dynamics (MD) simulations have been applied to shed light on the mechanisms of protein folding, conformational change, and protein-protein interactions. However, the computational time for large biomolecules, such as antibodies, is notoriously slow. This work combines high-throughput MD simulations and deep learning to develop surrogate models to predict antibody structural properties using only antibody sequences. These structural properties include solvent-accessible surface area, surface charge distribution, and surface hydrophobicity distribution. An extensive set of antibody sequences were retrieved from public databases. The homology models for these antibodies were built to run MD simulations and calculate the structural properties. The deep learning algorithm uses the antibody sequences as input and the structural properties as output for model training. Eventually, efficient surrogate models were developed to predict the structural properties of novel antibodies without the need to perform computationally expensive MD simulations. These surrogate models will facilitate the integration of computer simulations and data science for biomolecules.

585f Comparison of Dissociation Kinetics and Mechanism of Classical and Non-Classical Cannabinoids from Cannabinoid Receptors

Soumajit Dutta, Diwakar Shukla

Cannabinoid receptors (CBs) are part of the endocannabinoid signaling system, which help to maintain homeostasis in neuron signaling to control pain, obesity, and other neurological disorders. Therefore, synthetic cannabinoids (SCs) are designed and tested to target CBs as potential therapeutical selective drugs. Initially, SCs were designed by modulating the scaffolds of known phytocannabinoids. However, recent efforts have led to the rapid discovery of chemically diverse synthetic cannabinoids (Novel Psychoactive Substance; NPS), which have a high affinity towards CBs and significantly modulate the receptor activities. These molecules were started to get sold in the market as abusive drugs under different brand names (e.g., K2, spice) and caused thousands of hospitalizations of patients across the US due to more adverse effects, including impairment of fine motor skills and increased blood pressure, tachycardia. It is hypothesized that β-arrestin biased downstream signaling of these NPSs causes more adversarial effects compared to classical cannabinoids. How these classical and non-classical cannabinoids affect receptor conformational dynamics distinctly, has not been mechanistically studied. In this project, we compare the unbinding mechanism and kinetics of a non-classical cannabinoid, MDMB-Fubinaca, and a classical cannabinoid, HU-210, using biased and unbiased simulation. We implement transition-based reweighting analysis method (TRAM) to obtain unbiased thermodynamics and kinetics from the multi ensemble simulation data. This study provides a mechanistic understanding of biased signaling of non-classical cannabinoids.

585g Indacenodinaphthothiophene Isomers with Polyradical Character

Md Abdus Sabuj, Md Masrul Huda, Chinmoy Saha, Neeraj Rai

Polycyclic hydrocarbons with open-shell character are extensively studied due to their tunable molecular scaffold and electronic properties. However, design strategies delineating control of ground electronic state from closed-shell to open-shell diradical to polyradical character are not well studied. Here, we report indacenodinaphthothiophene isomers fused with five-membered carbonaceous rings and dicyanomethylene groups show large diradical and tetraradical character with small singlet-triplet energy gap, not observed in these polycyclic hydrocarbons. Density functional theory calculation indicates the *syn*- and *anti*-configurations have closed-shell ground-state with a large singlet-triplet energy gap. However, the *linear*-configuration displays pure open-shell diradical and tetraradical character with small hexaradical character. This study shows a novel design strategy of polycyclic hydrocarbons with a large polyradical character, which are compelling synthetic targets for magnetic and spintronics materials.

AUTHOR INDEX